Praise for
SACRED VIBRATIONS

"Exciting new studies show us that sound truly is medicine and can heal us in remarkable ways. Jeralyn's beautiful, heartfelt book, *Sacred Vibrations*, gives you an illuminating tour of this emerging discipline from one of its virtuoso practitioners. It is not a book of mere theory, though; it is a magical tapestry, woven with raw stories, inspiring examples, and practices that touch us at the core."
— **Stephen Dinan**, CEO of The Shift Network and author of *Sacred America, Sacred World*

"This book helped me to understand the healing properties and magic of music and sound, as well as giving me more understanding of the beautiful soul of the woman who wrote it."
— **Jada Pinkett Smith**, co-host of *Red Table Talk* and *New York Times* best-selling author of *Worthy*

"Jeralyn Glass is not just a master of sound, she is a master of heart. Her life's work, the depth of her lived experience and her mission to help others have made her work potent with transformation and a kind of inspiration that makes you approach your purpose with a little more reverence and joy!"
— **Devi Brown**, well-being educator and founder of Karma Bliss

"Within the vibrations and harmonies of sound and music there is magic! Music comes from the other side of the veil where there is a vastness of nothing but love. Your soul essence carries a remembrance of this music, which is filled with love and compassion. That remembrance is what Jeralyn Glass has been able to tap into to share the sacred vibrations of the universe with humanity."
— **Lee Carroll**, original channel for Kryon

"Have you ever been transported into another realm simply by listening to an exquisite symphony of music? Why is it that some can hear music from a certain composer and weep? Why is it that some individuals have even experienced miraculous healings because of certain harmonies and frequencies or specific melodies of music? The answers can be found within *Sacred Vibrations*, where Jeralyn Glass takes us on a journey of exploration to discover the miraculous and transformative power of sound and music."
— **Monika Muranyi**, spiritual teacher and author of the Kryon material

"A captivating exploration of the intersection between music, spirituality, and healing. Jeralyn's personal experiences are woven throughout, making the book all the more poignant. A sacred celebration of the sound vibrations connecting heaven and earth."
— **Anita Moorjani**, *New York Times* best-selling author of *Dying to Be Me*

"In this high note among books on sound healing, Jeralyn Glass takes the topic beyond theory and practice, making it both practical and deeply personal. Despite her unparalleled expertise in music and sound, she calls in a choir of other experts to round out this beautiful symphony of vibrational wisdom. Ultimately, this poignant teaching memoir is a primer for the soul's journey Home."
— **Suzanne Giesemann**, author of *Messages of Hope*

"With *Sacred Vibrations*, Jeralyn Glass combines moving personal stories with cutting-edge research and emerges as a leading voice in the field of sound healing. This book is certain to inspire and guide people in the use of music as medicine for healing and transformation."
— **Marci Shimoff**, happiness expert and *New York Times* best-selling author of *Happy for No Reason*

"The playwright Tennessee Williams once said, 'The world is violent and mercurial. It will have its way with you. We are saved only by love.' This same insight is everywhere in evidence in Jeralyn Glass's beautiful book *Sacred Vibrations*. She locates 'the frequency of love' in music and sound. She has a religious guide's grasp of the power of faith and forgiveness; a virtuoso musician's commitment to craft; a scientist's curiosity about why and under what conditions sound can heal us; and the heart of a great mother. Her story is a quest, not only for herself, but for us."
— **Mike Magee**, president of Minerva University

"Jeralyn's ability to convey the healing and transformative potential of sound is commendable. From a life immersed in music, she weaves together intuition, science, and knowledge of how vibrations affect our physical, emotional, mental, and spiritual health and well-being. This book is highly recommended for anyone interested in music for healing."
— **Karina Stewart**, chief wellness director and co-founder of Kamalaya Wellness Sanctuary

"This book warmed my soul! Jeralyn Glass is a treasured healer to this world. Her incredible teachings using sound and stories of her soulful son Dylan and their sacred connection are gifted to us in this beautiful new book. Pick up this book, read and hear the love it has to offer, and you will find a new undiscovered path to your inner knowing and healing!"
— **MaryAnn DiMarco**, psychic medium and author of *Medium Mentor*

"After spending the first half of her life filling up the world's stages with her mellifluous soprano, Jeralyn Glass is spending the second half of her life filling up the spiritual void inside us with the entrancing sounds of crystal healing bowls. Although scientists can't explain it yet, my most jaded and skeptical colleagues and I have witnessed the power of this new therapy to relax, reinvigorate, and repair even the most painful injuries through sound healing."
— **Dr. Daniel J. Levitin**, *New York Times* best-selling author of *This Is Your Brain on Music*

"Embark on a journey of self-discovery and healing with Jeralyn Glass's latest masterpiece, *Sacred Vibrations*. As a musician and seeker of intentional sound, Jeralyn weaves together the threads of her own experiences to offer a beacon of hope to all who seek mental, physical, and emotional balance. Through her profound understanding of the transformative power of sound, Jeralyn guides us to anchor ourselves in a system of wellness where intentional sound becomes our much-needed medicine. She shares how sound has been her steadfast companion, leading her from grief to joy. Join Jeralyn on this transformative journey and unlock the sacred vibrations within yourself."
— **Isaac Koren and Thorald Koren**, creators of The Songwriter's Journey

"*Sacred Vibrations* is truly a journey filled with Sound, Light, and Love! Jeralyn Glass shares her remarkable story of trauma, healing, and illumination. Her book is filled with extraordinary encounters with amazing beings throughout the planet who helped guide Jeralyn on her path of recovery after the untimely passing of her son Dylan. Through working with the vibrational essence of sound healing and in particular Crystal Bowls, Jeralyn has made a remarkable transition into health, wellness, and wholeness. There's much information on sound healing, crystal bowls, and much more. But most of all *Sacred Vibrations* is a heartfelt story of hope and redemption that will assist and inspire the life experience of every reader."
— **Andi and Jonathan Goldman**, authors of *The Humming Effect* and *Chakra Frequencies*

"A session with Jeralyn Glass is magic. I'm thrilled that her wisdom and relationship to sound is now bound into this book."
— **Gwyneth Paltrow**, founder and CEO of goop

"Jeralyn Glass's *Sacred Vibrations* is a transformative odyssey that harmonizes the soul's deepest grief with its most exalted joys. Through her personal journey, Jeralyn uncovers the healing power of crystalline sound and invites us to explore our own inner landscapes with courage and openness. Her narrative, rich and resonant, echoes the universal quest for peace, healing, and connection. As a composer deeply invested in the healing properties of music, sound, and vibration, I am moved by Jeralyn's exploration and application of sound as a bridge between worlds, both seen and unseen. Her work is a testament to the resilience of the human spirit and the transformative potential of music."
— **Barry Goldstein**, award-winning composer, producer, and best-selling author

"An authentic, high vibrational book about love, loss, sound healing, and the power of music to bring transformation and peace."
— **Lauren London**, actress

"Jeralyn Glass is a member of the legion of investigators pursuing in their various ways the use of music as an agent for physiological, psychological, and spiritual healing. On the other side of the transformational intensity of her life and musical experience, Dr. Glass absolutely radiates extraordinary joy and wisdom. *Sacred Vibrations* takes us through her own resonant story: an ongoing tale of music, life, exhilaration, despair, discovery, and healing."
— **Dr. Vern Falby**, former faculty at Peabody Conservatory of the Johns Hopkins University

"Musician Jeralyn Glass is wise to the ways of sound as healer. With tenderness, she guides us into vibrating singing bowls and music as medicine and miracle."
— **Joshua Leeds**, music producer and author of *The Power of Sound*

"Jeralyn Glass's *Sacred Vibrations* reads like a three voice Bach Fugue written inside a crystal. One voice sings to her son Dylan, another voice tells her story as a classical musician traveling the world, and the third voice is a crystal bowl singing. The voices sound an amazing counter point through which the reader enters the many facets of the crystalline sound and music of a singing bowl. Readers will discover, as I did, the voices in their own lives dissolving into heartfelt crystal clarity."
— **John Beaulieu, N.D., Ph.D.**, creator of Biosonic Tuning Forks and author of *Human Tuning*

"The Ancient Egyptians knew music as a way of interacting with the mind of God. This resonance with the Divine elevates human emotions to create the harmony of Heaven on Earth. Jeralyn has managed to elevate her emotional trauma to spiritual levels through musical experience and knowledge. This radiates through the crystal singing bowls. The BioGeometry team discovered they contain high levels of BG3, which is a harmonizing subtle energy quality present in all sacred spots around the world. This book unites the understanding of our ancestors with the wisdom of today through vibration and coherence."
— **Dr. Ibrahim Karim**, Founder of the Science of BioGeometry

"*Sacred Vibrations* is awe-inspiring! Jeralyn's life journey, and way of navigating tragedy, loss, and grief, producing the gift of *Sacred Vibrations* is testament to the power of frequency. This is truly the future of medicine, our resonant being, vibrating the cells and mitochondria and aligning our fields, is our true glory. Our ability to heal and become whole can be enhanced and expedited with the flow of information on these frequencies. Bravo!"
— **Dr. Greg Eckel**, founder of Energy4life Centers

"I am grateful that Jeralyn is exploring and exposing these ideas to the world! Her passion for healing with sound is invigorating and inspirational. I hope her book will ignite that spark within you! Thank you to Jeralyn and Hay House for bringing these concepts more into the mainstream and helping to shift consciousness!"
— **Maejor**, record producer, songwriter, and singer

"*Sacred Vibrations* is an absolute life-changing masterpiece that intertwines the realms of ancient wisdom and modern science and applies it to real life. Repeatedly, I heard from cancer patients who shared their stories of discovering comfort, healing, and hope through Jeralyn's knowledge and her crystal alchemy sound bath sessions. Through the subtle energies of crystals, they discovered a sanctuary within themselves, where pain dissolved, and tranquility reigned. This book serves as a roadmap, guiding both patients and caregivers alike toward a deeper understanding of the mind-body-spirit connection. I wholeheartedly recommend it to anyone embarking on their own journey of healing and self-discovery."
— **Nancy Lomibao**, Chief Clinical Officer of the Cancer Support Community, South Bay

"Jeralyn Glass's innovative work explores one of the most ancient human activities—the use of sound and vibration for healing—and supports its potential as a foundation of future research for the positive impact of the arts on health and the human experience."
— **Renee Fleming**, Grammy Award-winning soprano

"Blessings to all who read Jeralyn's book. The past three years have been some of the most challenging of my life. Every evening I listened to Jeralyn's astounding crystalline music on Youtube, and each and every evening a great sense of peace washed over me. And now Jeralyn has written the story of how an unthinkable tragedy placed her on the path to becoming one of the most powerful spiritual leaders in our world today. Her triumph over grief has taken her and those lucky enough to be in her presence to a higher form of consciousness."
— **Janet Leahy**, Emmy Award-winning television writer and executive producer

"Professor Glass' *Sacred Vibrations* is a tour-de-force exposition in the art of sound medicine, both ancient and modern, which beautifully and seamlessly marries sound with spirit. My healing experience in the Great Pyramid of Egypt, in 1997, led me on a quest to acquire knowledge of sound's therapeutic powers, and over a 25-year span, I have collected a significant library of books on this important subject. But during all those years, and in all those books, I have not encountered any that carry the loving heart energy contained in *Sacred Vibrations*. I highly recommended it for anyone on a path to discover sound as a catalyst for natural healing in the 21st century."
— **John Stuart Reid**, acoustics engineer and inventor of the CymaScope

"Jeralyn's words sing like music to longing hearts eager to meet the majesty of their own soul. A gift is found on every page and a treasure is revealed within."
— **Londrelle**, author, poet, and musician

"The combination of science and the sublime power of music to take us into the extraordinary; the pairing of Beethoven and crystal singing bowls, and a cast of leaders and practitioners in the field of sound medicine are woven throughout the pages of this informative and touching book. Jeralyn's collaborative journey brings us a symphony of "sacred vibrations." There are beautiful discoveries waiting for you. This book is for anyone wanting to explore sound as medicine, especially with the crystal singing bowls."
— **Anders Holte and Cacina Meadu**, musicians

"Jeralyn's faith in the healing power of sound is based on heart-wrenching personal experience, as well as a lifetime of dedication to serving beauty and joy to the world through the medium of music. Her story is key to understanding the depth of her life's work, and I'm so grateful that in this book not only does she elucidate compelling case studies and anecdotes from her professional practice, but also vulnerably shares her own journey of healing from the loss of a child, perhaps one of the greatest pains of life. Jeralyn is an ambassador for sacred sound. I pray the vibration and essence of this book reaches around the globe!"
— **Jahnavi Harrison**, musician and singer

"No one knows sound like Jeralyn. She is a master vocalist, musician, and sound healer. Her passion and knowledge of sacred vibration along with her ability to share and create healing spaces is an inspiration."
— **Jhené Aiko**, singer-songwriter and creator of Allel

"Jeralyn Glass is a brilliant educator and guide on the healing power of sound, and this book will open the hearts and minds of so many."
— **Lee Harris**, intuitive guide, singer-songwriter, and channeler

"I've been blessed to get to know Miss Jeralyn and the joy, healing, and clarity she's introduced into my life is insurmountable. By teaching me how to harness the power of my own voice and energy coupled with crystal bowls she introduced me to my inner healer. She's a true angel. May she reap all the love and beauty she sows in the world."
— **Solana Rowe (SZA)**, singer-songwriter

"Jeralyn Glass's *Sacred Vibrations* is a profound journey into the heart of sound healing, offering transformative insights and techniques for harnessing the power of crystalline sound and music. With her deep expertise and personal healing journey, Glass illuminates the path to harmony, wellness, and a deeper connection with the universal language of music. This book is an essential resource for anyone seeking to explore the healing potential of sound and its capacity to bring peace, balance, and joy into their lives."
— **Kristen Butler**, founder and CEO of Power of Positivity and best-selling author of *The 3 Minute Positivity Journal*

"Jeralyn and her son Dylan's journey inspire and teach us the power of love, sound, music, singing bowls, and mantra to bridge heaven and earth. In this book she has woven her inspiring story with the knowledge she has gained from the incredible synchronicities and guidance from her son on the other side, meeting teachers and experts in the field of sound healing and harmonizing modalities. As she healed her own grief, they, together in this informative book are bringing us closer to understanding and utilizing the magical mystery of sound."
— **Kevin James**, musician, flute player, and chant leader

"Jeralyn Glass, I bow to you with immense gratitude for birthing this book, which I know will bring great healing, sacred empowerment, and transformation to all who are blessed to read it at these wild times, and derive faith, courage, healing, creative intention, and self-remembrance."
— **Chloë Goodchild**, musician and founder of The Naked Voice

"I have been fortunate to attend Jeralyn's sound experiences, and they have been extremely powerful and sublime. In this book, Jeralyn shares this incredible experience with all readers, through masterfully demonstrating the transformative power of sound and music, and their ability to heal and awaken."
— **Jay Shetty**, host of *On Purpose* and *New York Times* best-selling author of *Think Like A Monk*

"Jeralyn is clearly part of a sacred army of love and enlightenment. And she adds to a canon of sacred texts with this deeply vulnerable yet enthralling revelation about the depths of music's power to heal. *Sacred Vibrations* is a timely prescription for so much of our world's current ills. The text transports you through the portal of Jeralyn's life into the transcendent possibilities of accessing a realm beyond our senses through music. It is both an exploration and an instruction manual for how to heal in a world so wrought with grief and tragedy."
— **Elaine Welteroth**, *New York Times* best-selling author of *More Than Enough* and **Jonathan Singletary**, musician

"Jeralyn has a beautiful way of speaking through vibrations. The vibrations of her singing bowls like the vibrations of this book leave you inspired, uplifted, and empowered to raise your vibration and live your most authentic life no matter what challenges you face."
— **Koya Webb**, author of *Let Your Fears Make You Fierce*

"*Sacred Vibrations* is a beautiful book about sound healing through which is woven the vulnerable and expansive story of Jeralyn Glass's transformational journey to wholeness through sound and vibrational energy. It beautifully combines the practical and the ineffable in a gripping read that will inspire readers into thinking of music as medicine."
— **Kulreet Chaudhary, M.D.**, neurologist, neuroscientist, and author of *Sound Medicine*

"Jeralyn Glass has composed one of the most in-depth books I have ever read on the healing power of music. She masterfully embodies the expression of the angels and reveals them so beautifully. *Sacred Vibrations* is brilliant, precise, and an incredible must read for healing your heart and soul."
— **James Van Praagh**, best-selling author of *Talking to Heaven*

"If you're ready to experience more peace in your heart, more joy in your life, and relaxation at a cellular level, look no further. In *Sound Vibrations*, Jeralyn Glass shares her wisdom, story, and the exquisite power of healing through sound so you can expand your vision, step in to grace, and truly transform your life."
— **Niyc Pidgeon**, positive psychologist and best-selling author of *Now Is Your Chance*

"Our pediatric hospice program has been blessed to experience the positive impact of crystal sound healing described in this book. Be it on Zoom or in person with Jeralyn or her Sacred Science of Sound Practitioners, they know how to transform sorrow into inner radiance."
— **Margaret Servin**, clinical supervisor for Providence Kids Care

"Jeralyn's expertise in *Sacred Vibrations* shines through in every page, offering a captivating blend of wisdom, science, and spiritual insight. This book is a true gem in the world of healing literature."
— **Lalah Delia**, author of *Vibrate Higher Daily*

"Understanding how vibrations and music affect our lives can help us enhance our lives. *Sacred Vibrations* is a book that can help us all."
— **Victor L. Wooten**, five-time Grammy award–winning bass player, songwriter, recording artist, and producer

"*Sacred Vibrations* is a generous and heartfelt journey deep into the heart of how sound has the power to transform, transmute, lift, and carry us through the hills and valleys of life. Moving, original and inspiring, this book is a must-read for any student of healing with sound and music."
— **Eileen Day McKusick**, founder of Biofield Tuning

"In her book, Jeralyn explains the curative power of sound vibration to improve and even heal a wide range of ailments and 'dis-ease,' from incurable cancer to the passing of a loved one. She includes scientific studies by expert doctors, Ph.D.s, and musicians to describe such minute details as the transformation and improvement of our very cell structure that occurs through the miracle of sound healing. This soothing modality of crystalline bowls has allowed thousands of parents and families to move forward and experience joy, knowing that our children in spirit are *still right here*. Thank you, Jeralyn."
— **Elizabeth Boisson**, president and co-founder of Helping Parents Heal, Inc.

"Music and sound vibration have been Jeralyn's life! From her extraordinary singing career around the world in the first half of her life, and following the devastating loss of her only child, Jeralyn has found healing for herself and a deep connection to her son through sound healing using crystal singing bowls. Her journey may help others struggling to find healing by connecting them to the medium of sound and music as a transformative medicine for the heart, body, mind, and soul. Even if a cure is not possible, healing is."
— **Dr. Glen Komatsu**, Chief Medical Officer of Providence Hospice & Palliative Care

SACRED
VIBRATIONS

ALSO BY JERALYN GLASS

*Crystal Sound Healing Oracle:
A 48-Card Deck and Guidebook*

The above is available at your local bookstore,
or may be ordered by visiting:

Hay House UK: www.hayhouse.co.uk
Hay House USA: www.hayhouse.com®
Hay House Australia: www.hayhouse.com.au
Hay House India: www.hayhouse.co.in

***

SACRED VIBRATIONS

THE TRANSFORMATIVE POWER OF CRYSTALLINE SOUND AND MUSIC

JERALYN GLASS

HAY HOUSE

Carlsbad, California • New York City
London • Sydney • New Delhi

Published in the United Kingdom by:
Hay House UK Ltd, The Sixth Floor, Watson House,
54 Baker Street, London W1U 7BU
Tel: +44 (0)20 3927 7290; www.hayhouse.co.uk

Published in the United States of America by:
Hay House LLC, PO Box 5100, Carlsbad, CA 92018-5100
Tel: (1) 760 431 7695 or (800) 654 5126; www.hayhouse.com

Published in Australia by:
Hay House Australia Publishing Pty Ltd,
18/36 Ralph St, Alexandria NSW 2015
Tel: (61) 2 9669 4299; www.hayhouse.com.au

Published in India by:
Hay House Publishers India, Muskaan Complex,
Plot No.3, B-2, Vasant Kunj, New Delhi 110 070
Tel: (91) 11 4176 1620; www.hayhouse.co.in

Text © Jeralyn Glass, 2024

The moral rights of the author have been asserted.

All rights reserved. No part of this book may be reproduced by any mechanical, photographic or electronic process, or in the form of a phonographic recording; nor may it be stored in a retrieval system, transmitted or otherwise be copied for public or private use, other than for 'fair use' as brief quotations embodied in articles and reviews, without prior written permission of the publisher.

The information given in this book should not be treated as a substitute for professional medical advice; always consult a medical practitioner. Any use of information in this book is at the reader's discretion and risk. Neither the author nor the publisher can be held responsible for any loss, claim or damage arising out of the use, or misuse, of the suggestions made, the failure to take medical advice or for any material on third-party websites.

A catalogue record for this book is available from the British Library.

Tradepaper ISBN: 978-1-78817-911-9
E-book ISBN: 978-1-4019-7271-4
Audiobook ISBN: 978-1-4019-7272-1

Cover design: Bryn Best
Interior design: Karim J. Garcia

This product uses responsibly sourced papers and/or recycled materials.
For more information, see www.hayhouse.co.uk.

Printed and bound by CPI (UK) Ltd, Croydon CR0 4YY

DISCLAIMER

Please note that the information in this book is not intended for diagnosis, treatment, cure, or prevention of any condition or disease. It is meant to assist you on your healing journey and does not replace medical care or therapy.

All content provided here is for informational purposes, and although we strive to provide accurate information, what we have presented here is not a substitute for professional advice regarding any issue, condition, or medical problem. Please note, as well, that some aspects of sound and vibrational healing are debated, even among the highly trained experts who are engaged in cutting-edge research and day-to-day practice in this field. While all attempts have been made to verify the information provided in this publication, neither the author nor the publisher assumes any responsibility for errors, omissions, or contrary interpretations of the subject matter herein. What you will find in the pages of this book represents the absolute best information the author has been able to acquire during her lengthy and expansive career and training in music and other vibrational pursuits. Some people may disagree with portions of what is shared here, but it is hoped that as the world becomes more accustomed to "music as medicine" there will be greater consensus on its intricacies.

No guarantees for medical results of any kind are being made by this author or publisher, nor are any liabilities being assumed. The reader is responsible for his or her own actions.

Before you begin any healthcare program or change your lifestyle in any way, please consult your physician or another licensed healthcare practitioner to ensure you are in good health and that the information contained in this book is suitable for you.

If you are experiencing severe anxiety and depression or an immediate crisis, please reach out to a mental health professional or crisis center or hotline.

About the Cover

The beautiful cover image shows the sound of a rare fifth octave F note made visible by acoustic-physics scientist, John Stuart Reid.

This photo was the result of playing a 6-inch Charcoal and Palladium crystal alchemy singing bowl. The sound created the sacred geometry of 22 elliptical islands called antinodes. The bowl is tuned to 432 Hz, and the F note corresponds to the heart chakra. However, with its high pitch, the note resonates in the realm above the crown chakra with celestial tones. The Charcoal alchemy helps to release stagnant energies, cleanse and detoxify the body, and clear the pathway between Heaven and Earth. The Palladium alchemy activates a state of awe and amplifies child-like wonder, playfulness, innocence, and joy.

The choice of this image on the cover has a deeper meaning: my son, Dylan, was born in Germany on January 9, 1996, at 22:22. One of his dreams was to be a professional photographer for *National Geographic* and at age 15 he won a prize for his photography. It is an honor to place the photo he took of a sunset on the cover of this book.

For my dearest Dylan: thank you for the joy and the wonder you bring, for the essence of pure Love that you are, for the peace you ground in me that goes beyond the mind and surpasses all understanding. I remember with awe the power of your birth at 22:22 in a small town near Bonn, Germany, on January 9, 1996. Sweet Soul of Light and Love, I had a purpose, then, as your mom. I have an even richer purpose now, as your mom. Eternal connection! With you always—forever Love.

My big, expansive boy loved the night sky and took many beautiful pictures of it. What a wondrous Soul to be a resident among the stars, to call their orbit his home. Together we have become the soaring one, the bridge traversing the broader, vaster, grander picture—him from the heavens, I from the earth. And we know; we love the stars so deeply, we have no fear of the night.

For my parents, who encouraged me to dream— and to follow my dreams.

Music is Love in search of a word.
— Sidney Lanier, *The Symphony*

CONTENTS

Foreword xxiii
Introduction xxvii

Chapter 1   **Grounding:** Music and Crystalline Sound to Anchor, Center, Elevate, and Transform 1
Chapter 2   **Birthing:** From Broadway to Berlin, Bonn, and Beethoven 17
Chapter 3   **Inspiration:** Kids4Kids and a Legacy of Love 31
Chapter 4   **Expansion:** A Bigger Purpose 43
Chapter 5   **Connection:** Sound, an Ancient Medicine 57
Chapter 6   **Amplification:** The Science of Sound Healing 69
Chapter 7   **Elevation:** The Quantum Is Real 83
Chapter 8   **Revelation:** Signs of Synchronicity in the Himalayas and Beyond 91
Chapter 9   **Integrity:** To Serve at the Highest Level 103
Chapter 10   **Curiosity:** Common Questions about Singing Bowls 113
Chapter 11   **Generosity:** Connections to Source 129
Chapter 12   **Tuning:** You Are the Instrument 139
Chapter 13   **Awareness:** Embodiment and Wholeness 153
Chapter 14   **Structure:** The Impact of the Crystal Singing Bowls on the Human System 163
Chapter 15   **Intuition:** Choosing Bowls That Resonate with You . . . 179
Chapter 16   **Possibilities:** The Ability to Reach Beyond the Visible . . 191
Chapter 17   **Vision:** Collaboration, Innovation, and the Future of Sound Medicine 205
Chapter 18   **Truth:** The Music Is Love 219

A Final Word 233
Endnotes 237
Glossary 239
Acknowledgments 243
About the Author 247

FOREWORD

Some people are born to explore, create, and offer their gifts to humanity for the benefit of all. They love to passionately discover and share what they find. Others blossom when they are exposed to these new perspectives. Sometimes the innovator doesn't have the opportunity to see what unfolds from their efforts in terms of development, awakening, or healing, but they share anyway because it's in their blood; it serves their soul, their dharma, their purpose. This is the case with Jeralyn Glass.

I first met Jeralyn on a JourneyAwake trip I was hosting to the Sacred Valley of Peru and Machu Picchu. She was deep in the devastation of having lost her son, Dylan, a few months earlier and was carrying out a dream of his, in his honor, to climb the mountains of Peru. She was shaken and broken wide open when she approached me to introduce herself. I sensed that this time for her would be life-changing indeed—not only in a healing sense with the horrible loss of her only child, but also in an evolutionary sense with the realization of an exalted experience of what the pain of love lost can become. I'd seen this in my clinical practice and felt its possibility energetically the moment she shared her story. The energy in the space between us literally held the promise of a great gift even though her words reflected her suffering. I told her she was in the right place. I felt she was destined for healing and something even beyond.

During our journey in Peru, Jeralyn declined a hike to the top of Huayna Picchu, as she was not up to the strain, given her emotional state. I took with me to the top a photo of Dylan to "show" him the view that he'd dreamed of as a child. In a sweet ceremony at the summit, I read his words inscribed on the back of his photo and showed him the 360-degree view of magnificence there. In that moment, I felt an amazing presence of Grace. One of the students took a picture to share with Jeralyn so that she too could witness this sacred moment. Later, on the train ride home, I saw that photo. There, one could clearly see a profound ray of light connecting Dylan's throat with the majesty of the great, blue sky—*the heavens*—as if to speak a great celebration of life. It was confirmation across the veil. He was there.

As a child, I discovered my ability to see the energy fields of the people and things around me. This gift came and went as I was influenced by my external environment through judgment and trauma versus support and care. Many people go through the same experience and fail to realize it on a conscious level until they begin to wake up to the truth of who they are. Upon my awakening in 1999 and 2000, my ability returned and stayed. What I had originally witnessed with Jeralyn when I first met her was a depleted state of energy due to her deeply painful circumstances. Later, I would see it transform into a beautiful, radiant presence.

As every great innovator does in the world, Jeralyn became a great researcher. She attended dozens of classes that I taught over the next few years. She was usually in the front row, daring me to explain how there could possibly be "good" in what had happened in her life as I shared about the universal principles of expansion and that everything is ultimately in support of our awakening. But the devastating loss of her son was consuming her process.

I would explain that everything was energy, including us, and that this energy was simply vibrating at different frequencies in the form of thoughts, emotions, and even our physical form. The combination of quantum science, bio-energetics, spiritual psychology, epigenetics, meditation, and BodyAwake Yoga was the foundation of her learning to build the circuitry to manage this time in her life from deep within, to retrieve the gifts these life experiences were truly bringing. Jeralyn's understanding of music, its theory, history, principles, and expression, allowed her to grasp that the power of an invisible energy—one that not only guides us but composes our very constitution—can heal us. She was able to do so, and even more beautifully, she generously shares her process here in the work that is before you. She has become more than curious; she is imparting new ideas, blending the worlds of her previous learnings as a professional musician and opera singer on an international stage with the holistic healing methods she has been exposed to. By doing so, she has become immersed in researching new ways of identifying their beneficial impact on personal transformation. She is now a leader in the field of sound and vibrational healing, using crystal bowls and voice transmission as her form of medicine. As with her own transformation, she sees profound results in the lives of the many who turn to her.

What science calls the quantum field, spirituality references as spirit and source. Both disciplines describe this essential foundation of existence as eternal and unwavering. Both claim that this invisible force is the ultimate power of life itself. The beauty of the new sciences is that they are pointing us in the direction of realizing that we have the power to guide, create, and heal through our conscious choices of learning to *identify as* this invisible force rather than merely live at the effect of it. We *are* energy, and "energy medicine," as it is referred to—which is any modality that uses this essence of life force to improve the flow and circulation of vital energy throughout our human system for healing—is something we can *each* utilize. Historically, sound healing is one of the original forms of healing.

Plato and Pythagoras felt the nature of the Soul was musical tone. Orpheus, born of Apollo and Calliope, was hailed as the first of the world's singers; his name means "he who heals by Sound and Light." Aristotle illuminated that as subtle sound waves pass through the density of the physical dimension, they generate our very life experience. While music is invisible, its impact is quite physical. Yes, as it turns out, the invisible world is the *real* world, and the physical world is its *effect*.

The subtleties are intimidating until they are not. We are here to refine our abilities to sense and feel—to perceive a higher realm of existence, a larger territory within which to play and create. I watched Jeralyn begin to master this as she utilized the processes that I teach. She began to cross through the veil that binds humanity and have deep and powerful connections with her son. I began to see, sense, and feel that the greater truth was emerging. They were dancing together once again, only this time in the invisible. Their exchanges were, and continue to be, most extraordinary. Jeralyn's energy field has become remarkably radiant, deeply heart-opening evidence of life beyond limits.

Science shows that energetic waveforms travel through the space between all particles and, in the case of the human system, hit upon receptor sites on the surface of every cell. This action transfers information to the interior of the cell and determines the chemical interactions the cell will initiate—it tells the cell what to do. These biochemical responses send neurochemical messages that rise through the body to the brain's sensory/motor integrative systems and determine the ultimate healing and harmonizing activities we experience. We have the capacity to manage the quality of these energetic transactions with the power of our own thoughts and feelings via the vibrations they emit. And we can utilize the beauty and bounty of nature through the use of modalities such as sound healing, vibrational training of the breath, meditation, polarity therapies, tuning forks, crystal bowls, and singing and chanting, among others, to help us align the chain of communication with our true nature of wholeness.

Sometimes we come to the moment of embracing these truths from a space of delightful and studious curiosity, sometimes for reprieve from a place of great pain. Whatever way we open, what matters is that we *do*. No matter how accomplished or how new to these ideas we are, there is always more to awaken to, actualize, and embody. Here Professor Jeralyn Glass will show you beautiful principles and practices to enhance the great healer within and to awaken your mind to the truth of who you are. Most significantly, she will do so with care, because that is who she is; more than curious, more than one who demonstrates how to play a crystal bowl, she is an innovator, a thought leader, a risk-taker, a passionate explorer, a pioneer, and a true healer.

Enjoy this magnificent journey.

With Great Love,
Dr. Sue Morter
Author of the best-selling book *The Energy Codes: The 7-Step System to Awaken Your Spirit, Heal Your Body, and Live Your Best Life*

INTRODUCTION

> Music is a higher revelation than all wisdom
> and philosophy. Music is the electrical soil in
> which the spirit lives, thinks, and invents.
>
> — LUDWIG VAN BEETHOVEN

From the time I was a little girl, I knew I would spend my life in music. I have a distinct memory of twirling around on our front porch singing and feeling the sense of freedom it brought, elevating me as if I had wings. I heard a voice vibrating within me that said, "Singing will be your life's path." And I befriended that voice throughout my life and followed it. It led me to New York, to Europe, and to the far corners of the world, where I explored the structure and emotional power of sound through different languages, cultures, and history, and I learned that the universal language that connects humanity *is* indeed the language of music. My trust and strength were birthed from that intuitive inner voice, which I understood to be something bigger than me—and yet *of* me. I named it Faith. It was my childlike, pure, innocent experience of God.

I knew early on that music holds the frequency of love, and love is the vibration that transforms everything. There is nothing love cannot heal. Little could I have known how my life was going to unfold and how potent that vibration of love expressed through music would become, a true guiding energy frequency in my evolution of consciousness.

In part, this book tells the story of music as the language of the heart, and how my son, Dylan, who was a strong, smart, and funny boy in life, became an even stronger, wiser, and humorous angel who shaped my deepened relationship with music as a powerful healing modality. He is collaborating with me on this journey as I explore the cutting edge in training, teaching, concertizing, and working with the crystal singing bowls as a medicine for these unprecedented times. The bowls have brought healing through heart-opening sounds and music to many people around the world, including me.

Dylan came into this world the way he wanted and left it the way he wanted. I had no power to change his choices. He was one of the most brilliant and entertaining people I ever met, and although our time together was short, it was incredibly full. I understand now it was not his fate to live into adulthood. During his life, he had six concussions while ski racing and playing football, which no doubt affected his ability to process his thoughts and feelings. At his lowest point, he chose to leave the planet. As a mom, I thought I could always be the miracle maker for my family, but I learned I could not change his destiny, nor what I came to realize was *our* destiny. Our souls had made a contract.

On the night Dylan died, my heart could not be stilled. I did not know what to do with myself. It was overwhelming. My family did not know how to console me. No one could help. I was in agony.

Peter, one of my dearest friends, came to stay with me. At midnight we walked down to the beach in Los Angeles to be comforted by nature. I needed the feeling of the wind on my skin and the smell of the ocean, and I needed to be bathed in the light of the night sky. Dylan and I had spent much of our short life together traveling; exploring in nature and discovering the world—from the Alps near where he was born in Germany to the mountains, oceans, lakes, and plains of Europe, North and South America, and the Far East. We skied, hiked, swam, rode bikes, and camped, and had incredible adventures we cherished. Peter, an actor, had also lived and worked in Germany, and he was on the board of directors of the children's foundation I created in Munich, the Kids4Kids World Foundation. He had known Dylan since my son had been a child. No words could express the pain I was in. The shock. The grief. Nothing, nothing had prepared me for this.

As we walked in silence, we looked up at the night sky simultaneously and saw a huge shooting star that seemed to begin right over my parents' home! Soaring high above the ocean, stretching in full glory across the Los Angeles basin, it arced to then land in the mountains that embrace the city. I heard a sound, a powerful *whoosh*, and I could hear Dylan's voice as if he were right next to me, speaking excitedly. "Mom, it's like we always talked about! I'm home. I'm with God!" This took my breath away. Did I just see my first-ever shooting star above Los Angeles, and did my son communicate through vibration, our favorite language?

A WONDROUS COMMUNICATION

Dylan was big—six feet, three inches tall and 230 pounds—and he had a strong, unmistakable voice. So lost in my grief, I could never have imagined what began that evening: a wondrous, new communication made of sound vibration and light that would open a relationship that has bridged the dimensions and crossed all known boundaries to show me that love is eternal. So much of what has been revealed through sound frequencies has opened me to a deeper life purpose. I have come to understand it is my responsibility to rise to the challenge and to *be* this unshakable love in the tidal wave of sadness. The only way through the anguish of this unfathomable void was to turn inward. The only way through unbearable grief was to wrap my arms around it and embrace it. My task was to master the acceptance of it, the breathing of it, the feeling of it, and find the courage to sit in it, transmuting it to joy.

Since that time the grief has been transformed into an inner radiance. Crystalline sound vibration centered in the magical grace of the crystal singing bowls carved the path. Dylan is always with me, anchored in my heart. And his presence is even stronger in his current form than it was in life. While Dylan was alive, it was the language of music, its ability to hold us in safety and comfort, and our ability to trust in this deep connection of our hearts and souls without any need for words, that made people often comment on our noticeable, uncommon closeness. We had a connection beyond this lifetime. Today, the gift of his communication through signs, frequency, and vibration has created the profound relationship we currently have, and the unexpected possibility of our miraculous connection through sound and energy has made all the difference. After Dylan's passing, as I began reflecting on how music and singing had affected my own childhood and teenage years, my adult life, and the 19 years we were blessed to spend together, I was astounded by how much the language of vibration and authentic expression were constant and guiding forces.

I was born and raised in Southern California to parents who had grown up in Brooklyn, New York, married at the young age of 20, and moved west. California brought them a new sense of freedom and possibility and provided a safe place to raise their growing family, which eventually included five children.

My dad would say to me, "The harder I work, the luckier I get." He taught me to dare to dream. He never told me to find a "practical" profession. He made a diagram on the yellow legal pad that was ever present in his life and showed me that there is a price to pay for everything; I needed to identify what that price was and then ask myself if I was willing to pay it. He and my mom supported me in developing the courage to go after my dreams with everything I had. I did. And I still do.

When I was 11, I sang a solo with the middle-school choir, and the response was overwhelming. Shortly after that my parents moved a piano into our home, and I began singing lessons with Annette Warren, a voice teacher with an extensive professional career. (At this writing she is 101 years old and still singing beautifully! Her wish—and mine—is that no one die with their song unsung.) She had been on Broadway, and she had dubbed actress Ava Gardner's voice in the MGM musical movie *Show Boat*. She had also dubbed for famous comic actress Lucille Ball. Annette was married to the great jazz pianist Paul Smith, who played for Ella Fitzgerald and Sammy Davis Jr., and she and her husband were raising their family near where we lived.

Annette gave me a powerful foundation in music and launched in me an enormous respect for the idea that our voice is a very personal instrument. She taught me a natural vocal technique based on the concept of wholeness, which included mental, physical, emotional, and spiritual awareness; I still teach it today. I include an expanded version of it in my sound healing trainings. Each of us has a unique vibrational signature that makes all the difference in our lives when we identify it, free it, strengthen it, and express it.

I attended college and then went to live in New York City, my eyes set on Broadway. The voice from my childhood that had whispered "Singing will be your life's path" continued to guide me.

Within a year, I received a Broadway contract to play the upstairs maid in the 25th-anniversary revival of *My Fair Lady*, starring Rex Harrison. I was the youngest cast member, and it was a dream come true! After one and a half years and 456 performances of *My Fair Lady*, my pianist suggested I explore classical music. I bought myself a standing-room ticket to the Franco Zeffirelli production of *La Bohème* at the Metropolitan Opera. It was so stirring, I knew I had to explore this kind of music. Funnily enough, as I transitioned to classical music, one of my first roles was Musetta in *La Bohème* a few years later and then Marzelline in Beethoven's only opera, *Fidelio*, which I would sing throughout my career.

Later I would find myself living in Bonn—Beethoven's birthplace—singing concerts and opera. Bad Honnef, a small village next to Bonn, is where my son, Dylan, was born. The music of Beethoven holds a place near and dear to my heart: when I was in the final month of my pregnancy, I sang a series of performances of Beethoven's Ninth Symphony and its "Ode to Joy." It is an epic piece of music for a full orchestra, four soloists, and a choir, and it embodies a message of peace and harmony—a vision of all people living in unity. Beethoven was an extraordinary human being who defied convention and broke his era's rules around what was considered "proper" music composition. His vision for music and humanity was vast, and I think his understanding of what it is to be a spirit in a human body must have been incredibly profound. He was able to communicate the most tender of human emotions in one moment and bring us the breathlessness of the angelic realms in the next. He brought us the thundering of hearts and the excitement of exploring new territory. And he did it all without the modern support of video images for his communication and, later in his life, without being able to hear a sound. It was as though Beethoven *became* the music he created as he poured the notes out onto the page. He knew something about music that his contemporaries did not. He was a man ahead of his time—perhaps even a man not of this dimension.

PROFOUND VIBRATIONS

Beethoven conducted the premiere of his Ninth Symphony in Vienna in 1824 when he had completely lost his hearing, experiencing his own masterpiece through its profound vibrations. Dylan heard this music in utero, so often that it was fully infused into his little, growing body as his mom rehearsed and sang the soprano solo nine performances in a row. Babies are sensitive to the sound signature of their mother's voice. It is their nourishment, bringing comfort, safety, and nurturing. After Dylan was born, I would play him this music whenever he cried. When I sang the soprano solo part, he would immediately become still. I was struck by the power of this musical balm that could soothe my baby's discomfort. It was very effective "magic medicine." Today one might calm a baby with a cartoon on a phone. For Dylan and me, it was Beethoven.

Dylan grew up immersed in the incredibly rich, diverse, and passionate world of classical music, traveling throughout Europe with me as I went from singing engagements in Belgium, Holland, France, Italy, Spain, Austria and Germany, including the former East Germany. Dylan was always with me until he began kindergarten, and he loved experiencing different cultures, languages, and places. The power of the human voice and great music of all genres fascinated us both. He loved to sing, play the piano and eventually the guitar. When he turned eight, we explored together the exquisite sounds of the crystal singing bowls, which he came to know well and use at a very vulnerable time in his life.

Introduction

It is a blessing to spend my life doing what I love, being a professional musician and performer, immersed in the teachings and the healing properties of sound for many, many years. In this book, I have brought together some of what I have learned over that time as a Broadway singer, then a concert and opera singer, then a professor and the initiator of a successful international children's foundation, and now the visionary behind the Sacred Science of Sound educational platform and the creator of the Crystal Cadence Sound Healing Studio and Temple of Alchemy. Everything I offer here relates to music as medicine, especially what I know and have experienced myself or through my many students around the world about crystal singing bowls, modern sound-healing instruments. I am privileged to share in this book the history of vibrational healing and some innovations in this field (with important musical concepts explained further in the glossary). I'm honored that a few dear friends and colleagues have so generously contributed their "sound" wisdom as well.

I invite you to experience this book in the light of knowing that vibration and music create the pathway and the bridge that can take us into the bigger picture of what is really happening in the scripts we write of our lives. They can show us what is truly possible for each of us, beyond those stories and the outwardly focused ego-based identities we have built. The expansiveness of crystalline sound has opened me to a bigger view of life *and* death and what I have held to be the truth. Dylan's passing changed my perspective on everything, deepening my experience of this human existence. I have come to understand our lives are a symphony of consonances and dissonances, sometimes harmonic and sometimes discordant. Most importantly, it is empowering to recognize that we are the creators of our lives: we are the composers, the conductors, and the singers of our unique songs, expressing our souls. Sound vibration helps us reawaken and rediscover this inner music—these melodies—and remember the greater orchestration and purpose of our lives. It guides us in quickening into the next evolution of who we are here to be.

I sometimes jokingly comment to my big Angel that a part of me does not remember signing any soul contract of this nature with him and that when I see him again, we are going to have a serious talk about it! I hear his cosmic belly laugh. Dylan and I had indeed made a promise long, long ago, and now in my consciousness, a distant memory of our soul contract lingers. We had agreed on a commitment to support people in awakening through the healing power of music, including the exquisite vibrations of the crystal bowls. In doing so, we are sharing a language we both love; bridging the worlds; helping people find their authentic expression, embody their Light, and live their life purpose with fulfillment and joy.

Joy.

No matter what our life circumstances, it is possible to come home to joy. For me to find my way to this deepened state of joy, Dylan had to go first, to be the bridge. That promise we made still takes my breath away. Through the life-changing work of Dr. Sue Morter, who has written the foreword for this book, I clearly remembered this agreement and the responsibility I have of living up to my commitment. Surprisingly, I discovered—with Dylan in his Spirit form—a much bigger reason for our relationship. Through the incredible power of energy, sound vibration, frequency and music, I received a true gift of eternal love that has defied the bonds of gravity, including time, space, and life itself.

I invite you to receive this miraculous story of courage, trust, and love, and to read it with an open heart and mind. I encourage you to breathe and feel all that is longing to be felt and then allow release, integrate all parts of yourself, and come home to the truth of *you*. There is no story or circumstance too big to keep you from your wholeness and your higher purpose. I am a living example walking that truth. Allow the sound meditations available through the QR code at the back of the book to ground you, strengthen your trust in the bigger picture, and activate higher consciousness. We are never alone. Please share with me a journey of wonder and awe, of healing and wholeness, through the transformative power of music and the sacred vibrations of crystal alchemy sound.

CHAPTER 1

GROUNDING

*Music and Crystalline Sound to Anchor,
Center, Elevate, and Transform*

There is a vitality, a life force, an energy, a quickening that is translated through you into action, and because there is only one of you in all of time, this expression is unique. And if you block it, it will never exist through any other medium and it will be lost. The world will not have it. It is not your business to determine how good it is nor how valuable nor how it compares with other expressions. It is your business to keep it yours clearly and directly, to keep the channel open. You do not even have to believe in yourself or your work. You have to keep yourself open and aware to the urges that motivate you. Keep the channel open.

— MARTHA GRAHAM, AS QUOTED IN *DANCE TO THE PIPER &
PROMENADE HOME* BY AGNES DE MILLE

It was so still, you could hear a pin drop. I nodded to the conductor, the orchestra sounded the first chord, and the energy of vibration began to move and amplify. As I stood on the stage of Europe's biggest cultural center, the renowned Munich Philharmonic's concert hall (*Philharmonie am Gasteig*), I felt the power of the music I was singing flow through my entire body and flood out into the audience in radiant waves of sound. Every cell of me resonated with the glorious notes of the famed "Vissi d'Arte" aria from the opera *Tosca* by Giacomo Puccini, and I felt the dazzling impact of being an instrument through which this vibrant music played to a sold-out crowd of 3,000 people. Music was the focus of my life's work, and what made the experience even more spectacular for me was the fact that this was one of the most celebrated events of the season—a New Year's Eve concert to ring in 2009—and I had been brought in to "save the show." The famous soprano who had been scheduled to sing had become ill, and the producer had called at 10:30 that morning to ask me to "jump in" and replace her. It was an incredible honor to be standing on that illustrious stage that evening!

During my classical music career in Europe, I was known as "the Jump-In Queen," which I attribute in part to the fact that my Broadway and national tour experiences, which had begun at a young age, had grounded me in an unwavering confidence that let me easily step into high-pressure performance situations. I had trained since my teens to be a vocal athlete, to practice with discipline, to show up fully present, and then to let it all go, trusting the energy of pure love expressed through music.

My motto? Take a risk, Jeralyn. Trust. Keep the channel open and then jump, and the net will appear.

The sky that morning was a bright, azure blue, and a fresh, blustery wind seethed around the city. But despite the cold, my insides were burning with anticipation. I knew exactly what I needed to do and what lay ahead for me in the next 12 hours. I began to prepare for the evening performance. I did 30 minutes of yoga to ground and open my body, then I began to play and meditate with my set of seven crystal alchemy singing bowls: one note for each energy center or chakra. Meditation with the crystalline instruments brought me an inner focus, balance, and clarity. I warmed up slowly, first humming with them; then, matching my voice to their resonance, I sang my vocal exercises as I played them. I felt strong. I showered, and under the flowing water I allowed my voice to siren up and down the scale from low to high and back again. After getting dressed, I reviewed the notes and words I was to sing and sat some moments in stillness with my eyes closed, visualizing the performance and setting

Grounding

my intention for a beautifully sung and inspired concert. I then packed my gowns, shoes, makeup, and a blue thermos of water containing fresh ginger, lemon, and honey—and drove to the concert hall.

My preparation routine served me well. There would be no rehearsal with the orchestra, as there was a matinee performance in the venue that afternoon that had gone overtime. This allowed me only a short time to talk through the music with the conductor and agree upon the tempi of the four arias I was to sing that evening. And it turned out to be one of the most extraordinary performances of my career. More importantly, it gave me one of many direct experiences of the transcendent power of music; I knew then, as I know so much more clearly today, that in those moments on stage, I was transmitting high-vibrational sound that was being powered by my breath; it steadily fed my two tiny vocal cords, expanding them and moving tones and melodies through my body into the concert hall. To project the human voice over the music being played by a large orchestra without any amplification is vocal athleticism. It was thrilling, and from the very core of my being, I knew that music and sound vibration have an extraordinary power to transform us.

Fast-forward to today, post pandemic. Science now calls this transformational quality an example of "music medicine." More people are experiencing the benefits of using the frequency and vibration of music as a prescription leading to mental, emotional, and physical stability; health; and wholeness. Globally we have been through unprecedented times, and life as we know it has shifted for everyone. Healing sound has become the "now" medicine as we seek to alleviate stress and pain and change the outward focus of our lives. We are learning the importance of an internal focus, of coming home. Sound helps us reunite with our authentic selves as we integrate mindfulness, self-care, relaxation, the setting of healthy boundaries, and regeneration into our lives. It is becoming an acknowledged solution for sustaining health and well-being.

SACRED CELEBRATIONS

Historically, we can see that music has always held an important place in our communities, accompanying our social interactions, our religions, and our cultures. It holds us in many of the sacred moments and celebrations of our lives. Music is the language of the heart, the language of harmony and unity. Quantum physics has proven that everything is energy, and everything is in motion.[1] On the physical level, our bodies are teeming with sound, termed biological or cellular noise—from our cells and organs to our skeletal structure and blood.[2] We are also crystalline in nature; our very bones are made of collagen and crystalline minerals; our cell membranes are liquid crystals and even our red blood cells have liquid crystal properties, all of which adds up to our human bodies being resonant quartz-like vessels reverberating with sound.[3] This is partly why the crystal singing bowls resonate and touch us effortlessly and why they have become the modern instruments of choice for transformation and for healing.

During my Munich Philharmonic jump-in, I knew I had to be in the moment, completely present, and *feel* in my body the sensations of that monumental experience, because they were anchoring and stabilizing something in me. I remember sensing the activation of what I would now name my Earth Star, or Legacy chakra, approximately two feet below my feet. I felt unshakable, grounded into the earth with roots well below my physical body. I was home and in my purpose. Time disappeared, and I was present to music alone in a connection that shimmered my senses into a pivotal experience of embodiment and upleveling. I felt a "quickening" that taught me to trust in a bigger way than I ever had before. That evening, I was intrinsically connected to higher consciousness, and it was flowing through me, inspired and unimpeded; I was soaring on the wings of elevated sound.

There is something magical about how music and sacred, intentional sound transports us beyond our thinking mind and invites us into the sublime. Great music gives an incredible gift, whether we are the listener hearing it or the artist creating it, and healing music is no different. When one *listens* to the singing bowls, there is a sense of being bathed in safety, perhaps feeling an opening of

Grounding

the heart and a quieting of the mind; in that stillness one may experience visions, inspirations, deep relaxation, and renewal. When one *plays* the singing bowls, there is an effortless sense of movement, creating a profound expression of transformational sound for yourself and for others. It is a glorious feeling of exchanging in a harmonic flow. As a singer and musician playing and producing healing sounds with the crystal bowls, I understand the theory and structure of music, and that is at the core of everything I play and every set of bowls I create for my students and clients. Crystal alchemy music is a powerful medium to integrate in our lives, especially now as we seek mental, physical, and emotional balance; health; and wellness. The pure frequencies and combinations of these astounding crystalline instruments help us ground, center, and expand. They support a graceful connection from the physical to the spiritual, allowing easy access to higher levels of consciousness. Indeed, the pristine sounds of the crystal singing bowls intensify this sense of the indescribable, despite the mathematical precision of the science behind their sounding.

The set of seven crystal singing bowls that I played regularly, including on the morning of the Philharmonic jump-in, supported an integral part of my preparation that day, as the pure quartz was anchoring to my system. Activating my voice in the way I did in my routine at home, then warming it up in the dressing room at the venue, primed my human instrument for its highest level of expression. I was excited and poised, ready to jump! At the time, I was also sharing the singing bowls with my young son, using them with my voice students, and bringing them into the kids' foundation program I created in Munich in 2006. The crystal singing bowls taught me the concept of entrainment and the possibilities that open within us when the bowls and the human voice attune to one another and resonate together. It was great fun for me, my son, and my singing students to match our voices to the crystalline instruments with all their rich overtones.

Though I had not yet formally studied sound healing, I was exploring vibration and frequencies, using the singing bowls purely as a musical instrument; and yet, miraculously, their sounds were already changing lives, including my own. The foundation for the path I presently walk was being laid. Playing the bowls was simple to master and, unlike with other instruments such as the piano or the violin, both a musician and a novice could bring forth glorious tones through them without hours of practice. They produced sounds unlike any other instrument I knew. I had loved playing the Himalayan metal singing bowls, but these crystal instruments were very different. My experience as a singer had shown me time and again that music was a medium that nourished our souls; little could I fathom that some years later, music and the crystal singing bowls would become a much-needed medicine for me that was to safeguard my life.

THE ETERNAL QUICKENING

Today, quantum science moves me to understand that there is and has always been a continual "quickening" translated into action through each of us. I received the Martha Graham quotation to Agnes DeMille that opens this chapter from a friend as I embarked on the national tour of *My Fair Lady*. I taped it on every one of my dressing-table mirrors as we performed across America and then returned to New York and played at the Gershwin Theatre on Broadway. I understood then the importance of allowing the essence that was unique to me to be fully expressed. My responsibility was to "keep the channel open." Now I would name this essence my vibrational signature and the channel the *sushumna*, the main energy channel of the subtle body uniting the chakras. This quickening is available to everyone, and as we allow this energy Graham writes about, this sacred, internal space to be felt within us, we will continually grow, expand, and evolve. We become so much more than our beliefs and our stories. We live our potential with an inexplicable grace.

Even more importantly, we can come to know that all that is happening is in support of the unfolding of our highest good. This attitude, and the recognition that we can choose both a deeper and higher viewpoint of life, serves in processing, accepting, and healing that which has wounded us. Love takes precedence. Our task, our responsibility, is to recognize our own uniqueness, brilliance, gifts, and talents and the personal "alchemy" of which we are made—our "vibrational signature"—and to give it permission to express itself. And this is all a part of my interpretation of the broader concept of music as medicine.

THE HEALING POWER OF MUSIC

Music has been acknowledged as a resource for healing since ancient times. In our own era, pioneering neuroscientist and *New York Times* best-selling author Dr. Daniel J. Levitin has collaborated with academic researcher Mona Lisa Chanda to evaluate the evidence that music improves health and well-being.[4] They have determined that music improves the body's immune response, reduces stress, and can be more effective than prescription drugs in reducing anxiety before surgery. Additionally, listening to music increases the body's production of the antibody immunoglobulin A as well as natural killer cells, the cells that attack invading viruses and boost the immune system's effectiveness. Music also reduces levels of the stress hormone cortisol.

The authors also note:

> In contemporary society, music continues to be used to promote health and well-being in clinical settings, such as for pain management, relaxation, psychotherapy, and personal growth. Although much of this clinical use of music is based on *ad hoc* or unproven methods, an emerging body of literature addresses evidence-based music interventions through peer-reviewed scientific experiments.

The authors propose music's stress-reducing capacity as a potential reason for its healing benefits, explaining:

> All organisms seek to maintain homeostasis. Stress can be defined as a neurochemical response to the loss of homeostatic equilibrium, motivating the organism to engage in activities that will restore it. Lifestyle choices that reduce stress are thought to be highly protective against diseases and music is among these.

It is stimulating to work with scientists and researchers in this field of music medicine to confirm what I experience regularly; music helps to calm stress, anxiety, and fear, establishing "homeostatic equilibrium" and easing physical pain, grief, and depression. This I know on a very personal level and have witnessed it time and again in both my own work with clients, that of my colleagues, and my students' work with healing sound vibration and crystalline music. My commitment is to continue exploring music as a lifestyle choice to support health and well-being. I received an invitation to speak and serve as an illustrative performing musician at the 2023 Sound Health Initiative workshop "Music as Medicine: The Science and Clinical Practice," co-organized and co-sponsored by the National Institutes of Health (NIH), the National Endowment for the Arts (NEA), the Renée Fleming Foundation, and the John F. Kennedy Center for the Performing Arts. It was an honor following the conference to be cited on the NIH website in an article called, "Sound Science: A Conversation with Drs. Simoni and Langevin." Dr. Hélène Langevin is director of the NIH's National Center for Complementary and Integrative Health (NCCIH). Dr. Jane Simoni is director of the Office of Behavioral and Social Sciences Research for the NIH. In the article Dr. Langevin praises the workshop and references my singing bowl presentation:

"First, the workshop highlighted the many examples of how music impacts the brain and the body. This dual aspect of music was vividly illustrated in the research studies that were presented, as well as the wonderful illustrative performances. I personally was struck by Jeralyn Glass' 'singing bowl' demonstration, which I found profoundly calming, both physically and mentally."[5]

Clearly, the more science and music collaborate, the bigger the win for humanity. In the words of Dr. Levitin, whether we play Bach, the Blues, bowls, or other genres:

> The music we choose for making us feel better is almost always more effective than that which is chosen by other people, regardless of their profession... it's often a good idea to have some sort of "Sonic Rx" on hand to use when we are feeling down and depressed. For some peo-

ple, this might be an upbeat popular song that they've heard when they were in a good state of mind, and which helps transport them back to that time of happiness. For others, it can be an extremely slow, sad song that reminds them that other people are in emotional pain as well, and that they're not alone.

Dr. Levitin jokes that an effective prescription for him to feel better is to "take two Joni Mitchell songs." For me, simply singing or playing bowls is all I need to do to feel better! The song that is my medicine presents itself in the moment and can be from a variety of genres.

As a classically trained musician and former university professor, I have the utmost regard for the science and theory behind how things work. But equally important are the surprising experiences that lead us to investigate what's going on out in the field. My team and I are gathering case studies of the sessions of the sound-healing practitioners around the world who have been trained through my educational platform, the Sacred Science of Sound. They are sharing anecdotal evidence that crystal alchemy singing bowl music is indeed a powerful healing tool for people of all ages, helping to maintain a sense of calm, promote sleep, reduce fatigue, anxiety, depression, and alleviate physical pain. It supports the processing of strong emotions, and brings a sense of self-empowerment in many health situations—such as with cancer patients, veterans with PTSD, hospice patients and their families. It is emerging as highly protective against disease because of its ability to reduce stress.[6,7] Some of my graduates are working with animals and the bowls, while others are playing in locations to heal the land where traumatic events have taken place. Throughout this book, you will read about the transformational experiences of several sound-healing practitioners and their clients.

I am passionate about what is developing with the crystal singing bowls' resonant music for the betterment of humanity. Important collaborations with leaders in the field are providing more proof supporting the power of sound and music to change lives. I am thrilled to share that we are working together with

Minerva University in San Francisco, ranked for the second consecutive year as the most innovative university in the world, creating a first-class Laboratory to research sound and psychology, the effects of crystal singing bowls, and music medicine. (For more on this exciting collaboration, please see Chapter 17.)

A number of innovative medical doctors are already using groundbreaking music medicine techniques in their practices. For example:

- The father of integrative medicine, Dr. Andrew Weil, and sound therapist Kimba Arem collaborated to create music meditations for self-healing.
- The late oncologist Dr. Mitchell L. Gaynor used both the Tibetan and crystal singing bowls in his pioneering work with cancer patients.
- Internal medicine and stem cell specialist Dr. Todd Ovokaitys uses special combinations of vowels and consonants for deeper levels of healing, and he also leads the renowned Lemurian Choir.
- Cardiologist, professor of medicine, and author Dr. Kavitha Chinnaiyan brings Eastern and Western medicine together, adding the element of sound.
- Neurologist, neuroscientist, and Ayurvedic specialist Dr. Kulreet Chaudhary integrates chakra toning and mantra meditation in her practice.
- Dr. Charles Limb, formerly at Johns Hopkins Hospital and currently UC San Francisco's Chief of Otology, is a hearing specialist and ear surgeon who runs the Sound and Music Perception neuroscience lab. He has centered his medical and surgical career around music, having been inspired by a deep obsession with sound, music, and the process by which our brains both perceive and produce it. A graduate of Harvard and Yale universities, he is now both a hearing restoration surgeon as well as a

researcher studying the neural substrates of musical creativity and music perception in individuals with cochlear implants.

Sound and Crystal Singing bowls are also an integral part of Energy4Life Centers under the guidance of Dr. Greg Eckel, where about 80 percent of the patients have Parkinson's Disease. Understanding that scientific inquiry is a longer-term responsibility, the Sacred Science of Sound and Crystal Cadence are committed to opening pathways to the scientific research of music medicine.

We are partnering with the Renée Fleming Foundation and the NeuroArts Blueprint Initiative called the Neuroarts Investigators Awards to encourage collaborative research between young and emerging scientists, artists, musicians, and researchers. Neuroarts is the study of how the arts measurably change the brain and body and how this knowledge is translated into practices that advance health and well-being. In this book, we will share anecdotal evidence of the powerful changes the crystalline instruments have activated for the people we serve through them. We are looking forward to all our collaborations and excited about the integration of crystal singing bowls into this field of music medicine.

> ### Sound Rx
>
> Wendy Leppard from South Africa is a graduate of my Sacred Science of Sound Crystal Alchemy training program, and recently reported a phenomenal exchange with a man who came to repair the gutters of her home. When the man arrived with his wife, Wendy invited them into her studio for a quick encounter with her bowls.
>
> "They sat down, and I talked them through a quick relaxation and connection into the field of the bowls," she says. "After playing the bowls for around five to seven minutes, I talked them back into being fully present in their bodies, at which point I instantly noticed a change in their demeanor. After opening his eyes, the contractor looked at his wife and said, 'The pain in my hand is gone.' Later, when my husband came home, he shook his hand."

> Wendy later learned that the contractor had had such bad gout and arthritis in his wrist that he had been unable to shake anyone's hand for ages. One short bowl session had healed his wrist and taken away his pain. While what happened was by no means unusual in our world, it is still a breathtaking example of how the crystal singing bowls in the hands of a skilled practitioner like Wendy can create a miracle.

THE HEALING POWER OF THE HUMAN VOICE: HUMMING RX

Fortunately, sound therapy, music therapy, music medicine, and sound healing are becoming more mainstream. Jonathan and Andi Goldman, authors of the award-winning book *The Humming Effect*, were a key part of my own path to healing with sound following the passing of my son, Dylan, and they have been training people in the targeted use of sound in healing for decades. Their expertise, experience, and vast knowledge are remarkable, and I treasure our friendship and collaboration, and especially their visionary research on the power inherent in our human instrument, our voice.

Everyone can experience the healing benefits of their own voice, as the Goldmans explain in their book. Everyone can hum, yet many people have been told they cannot sing. Having been a singer all my professional life, I feel it is important to remember that each of us is a unique vibratory vessel of sacred sound. No two people have the same vocal color and intonation, and this uniqueness of our personal frequency signature is what is being amplified through the singing bowls or any other sound-healing instrument we use. It is crucial to remember that no matter how beautiful a singing bowl may look and sound, we carry within us the most important instrument.

Jonathan Goldman shared with me that a lot of data has validated the therapeutic benefits of using our own voices, including humming, as a way to reduce pain, although the exact mechanism for pain reduction remains elusive. He proposed that it could work because the sound vibrations massage whatever part of the body is creating the pain, or the sound could be creating a resonance that negates and nullifies the vibrations of whatever is causing the painful imbalance. Humming also stimulates nitric oxide production, supporting healing in multiple ways, such as increasing blood flow and circulation, lowering blood pressure, decreasing inflammation, and boosting the immune system.

Another possibility Jonathan suggested is that self-created sound reduces stress on the nervous system and functions as a distraction therapy that takes a person's mind off their pain. (My dentist gets a kick out of me trying to sing and hum to calm my nerves during a dental procedure. For me, it works!) The healing capacity of our voice may result from a combination of some or all of these mechanisms, and the reasons for their effectiveness may vary from person to person. According to Jonathan, it is clear there is no one-size-fits-all recipe for success, but humming, singing, and listening to the human voice can help reduce pain, among other healing benefits.

Humming Your Way to Health

Jonathan and Andi Goldman have become great advocates of humming as a panacea for many different imbalances. Although they are experts in the use of all aspects of music—tones, frequencies, mantras, and sacred chant, for example—they have found that humming seems to be an all-inclusive sound. It's something everyone can do and it's something that everyone can experience as a therapeutic and beneficial aspect of sound. Refraining from judging oneself is an important aspect of humming that can bring transformation and a good dosage of the self-love medicine so needed by all. Jonathan and Andi confirm that our voices give us amazing abilities to create change. Scientific data on the beneficial physiological effect of conscious humming has shown it can[8]:

- Increase oxygen in our cells
- Lower blood pressure and heart rate
- Increase lymphatic circulation
- Increase levels of melatonin, enhancing sleep and easing depression

- Reduce levels of stress-related hormones such as cortisol
- Release endorphins—self-created opiates that work as natural pain relievers
- Boost production of interleukin-I, a protein associated with blood and plasma production
- Increase levels of nitric oxide (NO), a molecule associated with the promotion of healing
- Release oxytocin, the trust and kindness hormone associated with lowering stress and anxiety
- Increase heart rate variability, an indicator of increased resilience to stress
- Increase vagal tone, enhancing the parasympathetic nervous system and the body's ability to rest and digest
- Act as a vasodilator, which widens blood vessels and allows more oxygen to flow
- Enhance neuroplasticity, the nervous system's ability to build, repair, and create new neural pathways
- Function as an antiviral agent

A FORK IN THE HAND—OR EVEN TWO!

Tuning forks are being used by many leading sound researchers and therapists. The creator of Biofield Tuning, Eileen Day McKusick, treated me with her tuning forks shortly after the loss of Dylan. In Eileen's words, Biofield Tuning provides the opportunity to listen very deeply to the vibrational patterns in a person's electrical system, to hear and feel where a person is "at" without having to use words. This practice gives us a glimpse into a person's heart and soul, a window into their thoughts, feelings, and experiences.

"It gives people a chance to truly be seen and heard and felt, to be resonated with on a very deep level," says McKusick. "When I worked with Jeralyn for the first time, the massive grief that she was feeling from the recent loss of her son came through loud and clear, and we were able to be together in that grief without having to talk about it. This simple act of witnessing and holding space with someone, with what *is*—with no need to change it or fix it but rather just to *be* with it —can provide enormous relief and help us to integrate painful experiences that we might be challenged to do on our own."

McKusick relates, "The record of a person's experience of the emotions of sadness can be found in the biofield off the left shoulder. As I listened to this area in Jeralyn, I could hear the expression of deep, deep grief. This sounds almost like keening or wailing in the overtones of the tuning fork. Listening to the field is a way to really be with and empathize with someone who has experienced the depth and breadth of these kinds of emotions without them ever having to say a word. My heart ached right along with Jeralyn's in sympathetic resonance as I felt the unfathomable loss that she had experienced."

I was grateful for my experience with Eileen and Biofield Tuning, as it brought me a sense of safety and hope during a very unsettling, painful time in my life. The application of healing sound through my own voice, the tuning forks, and the crystal singing bowls helped stabilize my emotions and gave me strength to set boundaries and honor my healing process.

SOUND AS A TOOL FOR EMBRACING AND TRANSFORMING PAIN

So, whether we hum, sing, play the crystal bowls, work with tuning forks, or use any other sound-healing instrument, we have choices to support us in regulating our emotions and our pain. Our stories and our internal dialogue play an important role in our experiences of pain. We all struggle at various times in our lives, and pain of any kind can be overwhelming. We may forget that we are made of energy—that we are essence, Spirit in a human body—and as epigenetics has shown, we are so much more than our genetic inheritance and our family history; our beliefs and our environment affect our genes. Additionally, whether it is that unexpected, unwanted diagnosis, the loss of a job or opportunity, a shift in a relationship, an accident or death, or a financial crisis, these events are all a part of being human. We are not victims of circumstance, especially if we choose the attitude that the occurences of our lives help us to grow, awaken, and evolve.

Yes, it is uncomfortable to sit in our pain; yet when we acknowledge it, breathe, and feel it, pain can be our greatest teacher. We are taught to get rid of it as soon as possible; whatever it takes, we want to avoid it and eliminate it. Pain revealed to me a bigger purpose, and pain empowered me. I never dreamed that losing my son could hold anything positive for my life; yet working with breath, energy, and healing sound showed me an unexpected and miraculous pathway that I am honored to share with you.

Energy and music medicine are fields with limitless possibilities, and it is exciting to be an integral part of this exploration. Music and intentional sound anchor us, quieting the inner chatter of our mind, transporting us beyond where words, human touch, and even medicines can go, into the realm of the sublime, the exalted. When we land there, we connect with our Soul. In this sacred space, possibilities abound, and healing is real.

I am still astounded by what has transpired within me. I never imagined I could feel happiness again, and yet joy resonates within me. And my son is ever present through energy, vibration, Light, and the grounding and elevating frequencies of the singing bowls. I am so glad you have joined me on this incredible journey as we explore vibrational healing, crystalline sound, and music medicine, and discover together how these may also be helpful for you.

CHAPTER 2

BIRTHING

*From Broadway to Berlin,
Bonn, and Beethoven*

I am of the opinion that my life belongs to the whole community, and as long as I live, it is my privilege to do for it whatever I can. I want to be thoroughly used up when I die, for the harder I work, the more I live. Life is no "brief candle" to me. It is a sort of splendid torch which I have got hold of for a moment, and I want to make it burn as brightly as possible before handing it on to the future generations.

— GEORGE BERNARD SHAW

With the blessings of my parents, at the young age of 20, I left California to follow my dream of being a professional singer in New York, and my path revealed itself rather quickly. In reflection, now I see how everything that unfolded in my life and music career was a preparation for utilizing sound as medicine, leading me to the crystal bowls and the dimensions of heightened awareness for which they are such a perfect portal.

My first audition in New York was for the role of Mary Magdalene in a regional theater production of *Jesus Christ Superstar*. I sang "I Don't Know How to Love Him" and was hired on the spot. A dream had come true! In retrospect, it's no coincidence my professional debut was as Mary, and I felt humbled singing at the feet of the actor portraying Jesus. I experienced a feeling of unconditional love that would serve me at the time of my deepest sorrow. I understood I was playing a prostitute, and yet never felt that was the truth of Mary: her love felt so pure. Later I learned that the Catholic Church had decreed Mary was not a prostitute and historical references concluded she was from a wealthy family and was a beloved companion and teacher to Jesus and the apostles.

Mary Magdalene expressed profound wisdom through the energies of the divine feminine. She wore red because it represented the grounding of devotional and exalted love. Red energy is crucial to our ability to expand, integrate, and sustain. Red is the color of the root chakra, which is associated with stability, safety, and belonging. It was a perfect role for my professional debut and later, as I would hold my son in my arms, both at his birth and his death, I would understand this had been a dress rehearsal.

After Jesus Christ Superstar closed, I studied acting with master teacher William Esper, who taught me about presence, how to be truthful and authentic in action and responses, and how to create a character. I began regular auditions, which was an art unto itself.

The Audition Process: Eight Bars, Please!

I remember the day I waited hours in line to sing an eight-bar chorus audition for a national tour of Jerome Kern's musical from 1927: *Show Boat*. At these chorus-call auditions, we had to count our bars of music before we sang. (For a reference point, eight bars can sometimes be only 15 seconds long!) I had to acclimate myself to having an eight-bar window when I started auditioning for big shows; there was no time to sing a two-minute song. The audition team knew if they were interested or not within a few seconds of a singer walking into the room and opening their mouth.

I got good at those quick auditions. I was hired for *Show Boat*. We played at the National Theatre in Washington, DC, and Wolf Trap National Park for the Performing Arts, my first outside amphitheater experience. (Help! what do you do when you are singing and a bug flies in your mouth? Swallow it or spit it out discreetly?) A joyous life of music and travel had begun.

Birthing

After a successful audition for the national tour of *Showboat*, I traveled the country blissfully performing, yet all too soon the run was over, and auditioning began again. I was thrilled when I received my first Broadway contract playing the upstairs maid in the 25th-anniversary Broadway revival of Alan Jay Lerner and Frederick Loewe's *My Fair Lady*. The original lead male actor, Rex Harrison (an English stage and film actor, one of Hollywood's then leading movie stars) and the original actress who played his mother, Cathleen Nesbitt (aged 92 at the time!) were also in that production.

I am forever grateful for the internal growth those shows catalyzed. We performed five evening performances a week, with a matinee performance every Wednesday, Saturday, and Sunday. I loved my work! I learned to understand and live in the "quickening" demanded of a high-performance career. Dad always reminded me, "Do what you love and love what you do. Then it is never work." I was constantly inspired. During these beginning years, I learned how to immerse myself in a character and memorize the music, lyrics, staging, and choreography while understanding that the task was to recreate every performance as if it were the first one. Every single performance had to have spontaneity and freshness. That is the art and craft of live theater. And still today I sing, play, and teach following the same guidelines I learned as a young professional. Although I may play the same set of alchemy singing bowls, it is important to be grounded, present, and a clear vessel willing to create anew in each and every moment. No two sound-bath experiences—immersive listening sessions in which attendees are bathed in sound waves—are ever the same, even with the same set of bowls, just as no two performances are ever the same even though it's the same show.

> ### Surprise! Grandma Ree Dear
>
> One of my great memories of the *My Fair Lady* tour I did with Rex Harrison took place during our three-month sojourn at the Pantages Theatre in Los Angeles. I invited him to my parents' home for lunch, and he accepted. We were excited and decided to keep it on the down-low even with my family members. However, my barely five-foot, one-inch grandmother—Marie, affectionately known as Ree Dear—was taking a Monday drive and decided to drop in unexpectedly to visit us. She rang the doorbell and proceeded to march up the stairs. She took one look at Rex Harrison sitting on the sofa and screamed, "Oh my God! That's Rex Harrison! I think I'm going to drop dead!"
>
> Rex did not skip a beat. In his most elegant, gentlemanly manner, he stood up and said, "Oh please, madam, don't do that!" We all laughed, and what followed was a most charming conversation between Rex and my grandmother. She was starstruck to see him at her daughter and son-in law's house, although she had already been twice to see the production at the Pantages Theatre. Afterward, she called and asked why she had not been invited. I believe today it was exactly so the universe could bring her to the party in a more humorous fashion that we all never forgot. We laughed about that day for a long time. Ree Dear forgave me, and I came to recognize the importance of trusting the bigger picture: she was obviously meant to meet Rex, whom she adored from the silver screen, no matter what I had thought!

After *My Fair Lady*, I made it to the final round of auditions for several major Broadway productions but did not land the part. Although the reason was not yet clear, musical theater doors were closing while a new pathway was opening. I followed my inner guidance, and when the music of the classical world began calling, my career quickly shifted gears, as it was supposed to. The path from Broadway to classical music evolved naturally, and I went to study for a year in West Berlin. It was an exceptional time in history when Germany was divided into East and West, and Berlin was a city of four sectors: the American, Soviet, British, and French. Living there gave me experiences that further shaped my perspective of the power of music to overcome differences and unite us. I have since sung concerts and played singing bowls in places as far and wide as Tonga, Chile, Mexico, Cambodia, Thailand, Bhutan, Indonesia, Australia, Singapore, China, Japan, and throughout Europe, where spoken language was a barrier and music the common language, yet I will always remember one unbelievable night as a young singer at the beginning of my opera career living in a divided Germany and how it continues to influence all that I do in the world of music and sound healing.

My hands were gripping the steering wheel of my borrowed Audi 80 as I sped back along the autobahn to West Berlin from my singing teacher's home in Hamburg. These were tense days politically: the Cold War was raging, and I had lost track of how many East Germans had been shot trying to escape to the West. But although it was pitch black and the hours were inching toward midnight, I was oblivious to darkness of any kind. I had just spent the day training in Hamburg with the great American soprano Reri Grist. I had begun the four-hour drive back to West Berlin full of energy and inspiration, and the music from my cassette tape danced through the car speakers.

I was singing along with it, in my happy zone, when I suddenly realized I was lost. There was darkness all around me, and I couldn't see any buildings. The road had become narrow and bumpy. I slowed my vehicle and took a deep breath. Then panic flooded through me. My heart beat faster. I was frightened. Nothing looked like what I remembered from my excited drive west to Hamburg earlier in the day. What had happened? I drove slowly until I saw a distant flicker of light to my right. Approaching it, I saw it was a *Gasthof*, a kind of casual restaurant very unlike the diners I had frequented while living in New York. This restaurant looked somber, gray, and unwelcoming. I was a stranger here, a young American woman living in Europe. I parked, got out of the car, steadied my wobbling legs, and made my way to the entrance.

A rather stout middle-aged woman greeted me brusquely.

"Excuse me," I said in my best German: *"Entschuldigen Sie bitte.* I was looking for the exit to Berlin. . . and none of this looks familiar. I think I'm lost," I finished. "Can you help?"

The restaurant was full, and the patrons looked at me like I was from another planet. For them, I was. I had landed in a small town in East Germany, close to midnight, and it was to be hours before I would arrive at the tiny apartment in West Berlin that was then my home.

"How did you get here?" the woman asked me sternly. She leaned toward me with a shaming posture.

A tall, thin man, his tone toxic, joined in. "Do you have a permit? You are not allowed to be here without *ein Erlaubnis*."

I was trembling.

"How do I find the way to West Berlin?" I asked meekly. I understood I was in trouble, perhaps even in real danger. As an outsider in a foreign land ruled by Communism and a strong military government, a mistake like this could have big consequences. I listened as best I could to the directions they gave me and jumped back in my car.

A COUNTRY DIVIDED BY POLITICS

I buckled myself in and followed the instructions back into the shadows of a country divided by politics and populated by a common people who had been separated by war, by ideals, by dictatorship, and by fear since 1949.

How could I have been so unaware?

I realized then that I had taken the exit marked *Berlin Hauptstadt der DDR*, (Berlin, capital city of the German Democratic Republic) which led to the heart of Communist Germany. I should have taken the exit marked *Berlin Bundesrepublik Deutschland BRD*, (Berlin, Federal Republic of Germany) for West Germany.

Ohhhh Toto, I thought, (referring to Dorothy and her dog Toto from *The Wizard of Oz*, my favorite film as a child) we are definitely no longer in Kansas. As I got back onto the autobahn—one of the main superhighways that runs through Germany—I let out a huge sigh of relief and my nerves began to calm... until, looming above me, I saw one of the big border-control structures that dotted the highway. My heart sank once again.

Birthing

The border guards were in uniform, and they were heavily armed. Gruffly, they asked me to pull over and get out of the car. I was frightened. As I stood outside in the freezing-cold winter night, I watched as they removed every single item from the car, including each one of my cassette tapes. I politely answered their questions.

"I am a singer, and these are recordings of my vocal exercises and the music I'm practicing." They looked menacingly at me and proceeded to take apart every piece of the upholstery of the Audi. One of the guards lay on the ground and wiggled underneath the car, then checked with his flashlight to see if I was smuggling anything; another looked in the trunk to see if anyone was hiding there. I don't know which made my trembling worse, the bitter cold or my fear, as I watched them check every millimeter of the vehicle with their flashlights.

Will they arrest me, and I'll never be heard from again?

At the end of their investigation, the guard in charge of checking papers looked at my passport, then at the clock, and said in German, "Miss Glass, happy birthday!"

His wishes startled me until I registered that it was well after midnight. And it was now December 7, which was indeed my birthday. My eyes filled with tears.

I replied, *"Danke schön."* Thank you.

These men were doing their job, guarding their borders. A light of compassion turned on inside of me and renewed my faith in humanity. The fear was gone. I remembered the stories of brutality that had been meted out as the Berlin Wall had been built and how families and friends had become divided. The officer had made an eye-to-eye connection that calmed me. It was 4 A.M. by the time I got back to my apartment in West Berlin.

Today it is hard to fathom what life was like then in Germany, a country mercilessly divided: the armed borders, an impenetrable wall around the city of Berlin. People who tried to reach the West were arrested, deemed traitors, and imprisoned. . . or simply shot in cold blood. Media and television were restricted, and parents were spied on and fined or even imprisoned if their children watched Western cartoons. This seems unimaginable; yet it was true from 1961 to 1989. Europe as we know it today did not exist, and many people were not free to move about. During the year I went to live and work in a divided Germany, my understanding of the meaning of freedom deepened, and I realized how many simple things, such as going to the movies or grocery shopping, I took for granted. Years later, after the wall fell, Germany was reunified, and my classical music career took off. I came back to Europe, choosing Bonn as my home base. I performed often in the former East Germany and experienced music time and again as the great unifier. Regardless of our differences, music creates a vibrational connection between human hearts where beliefs, nationality, culture, language, religion, and social status do not matter.

And there, in the heart of a newly reunited Germany, I became the mother of the beloved boy who would radically change the trajectory of my entire life. This book would never have been written without the birth of my son, Dylan Sage.

AN EXCITING YEAR

I remember how ecstatic I was when I found out I was pregnant! It had been an exciting year of singing engagements, and the last one had been a jump-in in France where I learned a complete operatic role on my feet in a week, musical score in my hand during staging rehearsals, and then performed it eight times. It was the most intense experience of my career up until that point. My body was so weary, it could not resist any longer; it opened and received a new life.

The pregnancy was easy until my son arrived two weeks early. I had been singing a run of Beethoven's Ninth Symphony, with its rousing last movement for soloists and choir, "Ode to Joy," very pregnant, in great form, with plans to stop well before the baby's due date. It was complete bliss to be nourishing this little being inside me and singing music I adored. On the evening of January 8, I drove from Bonn to Köln to the Kölner Philharmonie concert hall to hear a friend singing a performance with Sir Neville Marriner conducting, the maestro under whose baton I had made my Los Angeles Opera debut. Driving home, I began to get labor pains. What began as a planned home birth in front of the fireplace ended up some hours later in the hospital: there were complications. It became a long and trying labor, but finally, on January 9 at 22:22, my child was born near Bonn in Bad Honnef, Germany.

I will never forget that first moment when the medical staff handed him to me and our eyes met. It felt as though we had been together before. Our eyes locked in the recognition of the soul, a look of eternal knowing, and an abiding love. In that moment of a magical, vibrational connection, I heard his delight and humor, imagining him laughing and commenting, "Ah, Mom! So that's what you look like this time around."

Gazing intently into the eyes of my newborn child, I silently responded, laughing, "Oh yes, son, and this is who *you* are this time around. Ha!"

It all felt so familiar. I was ecstatic to meet again. Our bond was eternal and beyond words from the first moment; I loved my son with every fiber of my being.

I wanted a name that could be pronounced the same in both German and English and chose Dylan, which means "born of the sea." And a water baby he was. When I was six months pregnant, I had swum with wild dolphins off the Florida coast. A pod consisting of three adult dolphins and two babies had swum around me. I felt them communicating with the little one growing inside me. There was a connection with these beautiful mammals through sound vibration, and this was something Dylan and I would share throughout the short time we had together.

In Germany, I had the option to take time off from my contract at the Opera House to care for my newborn and still be on salary. I took that year to be with my son, although I did sing a few jump-ins. The first one was when Dylan was only four months old, to perform in Mozart's *Così fan tutte* for the opera house in Augsburg about four hours south of Bonn. I had sung *Così* in three languages—English, German, and its original Italian—and this production was in Italian. It would be my first performance since Dylan's birth. I packed our bags and we hopped on the train, Dylan happily nestled in his little carrier. When I arrived at the theater with him, I was fitted for my costumes and went straight to rehearse the music with the conductor and cast, then directly to the staging rehearsal.

The following evening was the performance. In the opera world, a stage director and designer shape the look and feel of a production, just like the conductor shapes how the orchestra plays the notes and phrases. The artistic team puts a "signature" on the performances, colored even further by the leading singers. This *Così fan tutte* was a modern production, and my costume as the maid Despina was a skin-tight, sequined minidress with tall, thigh-high leather boots, a huge beehive wig, and heavy modern makeup. Baby Dylan was with me in my dressing room, mostly sleeping in his carrier, and the assistant looked after him when I was onstage. When it was time to nurse, he did not recognize his mom and gave me an indignant look that said, "Like. . . really? Who are you, and where's my mom?" I had to laugh. He looked at me suspiciously, and he began to cry uncontrollably. "Dylan, I reassured him, it's Mommy. It's okay." As he heard my voice, the vibration he knew so well, he calmed down. As I started to nurse him, all was familiar; he knew he was safe and sound.

Dylan traveled with me to every singing engagement throughout Europe until he started his formal education. There was Venice and Teatro La Fenice, where a pigeon landed on Dylan's head in Saint Mark's Square. He learned to say, "*Buongiorno. Mi chiamano Dylan.*" (Good day, my name is Dylan), and we enjoyed a charming serenade in a gondola. I sang at the Opéra de Marseille, and we explored the southern coast of France, finding a little seahorse that would become so meaningful to me many years later, causing in me a profound quickening. Dylan learned to say, "*Bonjour, madame, je m'appelle Dylan. Je voudrais un jus de pomme.*" (Good day, madam, my name is Dylan, and I would like an apple juice.) And then came the opera in Málaga. He splashed in every fountain in that city, discovering a new joy in his element—water. I can still hear him say, "*Hola, me llamo Dylan.*" (Hi, my name is Dylan.) I went on tour throughout Holland in Mozart's *Abduction from the Seraglio,* and we learned some Dutch: "*Hecke becke trecke,*" which means "making funny faces" (which he loved to do!), was Dylan's favorite phrase. Every season I had at least one engagement in France, where Dylan collected all 16 *Tintin* comic books by Hergé in both German and English. He loved the leading character of Tintin and his little dog, Snowy. Once he was in school and could not join me, I would write him postcards daily, and he would write me back. We accumulated quite a collection of correspondence that later, after his passing, would reveal an eye-opening view of our soul contract. And we still managed to see the world when school didn't keep him occupied.

We traveled to the United States every year during my summer break from the opera to see my family, and when Dylan was barely four years old, we went on a day excursion in a small boat to swim with wild spinner dolphins off the Kona coast of the Big Island of Hawaii. It was a radiant day, a perfectly blue sky, a clear and crisp horizon, and a sun that danced its beams upon the sparkling, intensely colored turquoise ocean. I strapped Dylan up in his little life vest, and a new adventure had begun. We swam with curiosity and joy, flipping our fins and looking through our snorkeling masks, watching hundreds of spinner dolphins swimming below us. Sometimes the dolphins jumped up in the air and spun around right in front of us. It was a phenomenal sight. As the huge pod of dolphins swam on ahead, we surfaced, removed our masks, and looked at the vast stretch of the water in front of us and all around us, speechless. I saw Dylan's eyes take serious note that he was here, in the middle of the Pacific Ocean, held up by nothing but his life jacket, and he could not yet swim! His face registered it all in the blink of an eye: fear and exhilaration were superseded by a decision. He was done. "Mommy," he said, "I'm going back in the boat."

I laughed and climbed into the boat with him.

TRAVELS WITH DYLAN

Dylan made friends wherever we went, human and otherwise. We were blessed to visit South America with my mom, and we saw iguanas of all shapes, colors, and sizes, which he loved! He connected with a small seal he called "Nibbles" in the Galapagos Islands. It frolicked with him and chewed at his snorkeling fins. And there was a horse who carried him high up in the mountains; he cried when they had to say goodbye. This journey was one of the most fulfilling experiences of our lives: we swam with a hammerhead shark, discovered blue- and red-footed boobies, saw a turtle that Charles Darwin had also seen, and we watched some turtles procreate. That came with some highly unusual sounds we had never heard before. We laughed so hard! Those were certainly high vibrations!

Dylan loved exploring the world with me. He loved the seasons and nature, he loved God, he loved music, he loved culture and sports, and he loved life. When he was very young and I would put him to bed, we would pray, express our gratitude for the day, and then do a little meditation together. We would close our evening with him saying, "Mommy, I love you as big as the mountains." (We lived in Bavaria in the south of Germany bordering the Alps.)

I would say, "Sunny"—one of his nicknames—"I love you as big as the moon."

"Mommy, I love you as big as the stars."

"Dylan, I love you as big as the sun."

"Mommy, I love you as big as all the planets."

"Sunny, I love you as big as the universe."

And then he would top me by saying, "Mommy, I love you as big as God, and there is nothing bigger than that!" He always had the last word!

Now I say to him, "Dyl, I love you as big as God, and... I love you beyond the beyond."

He still has the last word, and he whispers as a vibrational frequency, "Mom, I love you to Eternity and Infinity, and that never ends!"

FINDING THE BOWLS

On one of our visits to the United States, my mother and I made time for a girls' getaway. I saw an announcement at our hotel offering a crystal singing bowl presentation that evening and was intrigued. I had a collection of the Himalayan singing bowls that I used regularly and were part of my meditation practice, but I had never heard crystal singing bowls. I asked my mom if she would like to go. She is open to learning new things and of course said yes. Together, we attended the offering and discovered a new instrument.

When I heard the pristine sounds of the crystal singing bowls, I was transported. They were exquisite and resonated deep within me. The vibrations were familiar, like a song I knew from a distant time. I had to purchase some of those bowls! I bought seven with different alchemies and different tones, each note corresponding to one of the seven chakras, and I brought them back to Germany, where I shared them with my family, my students, and my friends. Everyone loved the sounds.

Dylan was seven when I brought these phenomenal instruments home, and he absolutely adored them. He'd ask me to bring him to bed with his "sound blanket." The bowls then became a part of our expanding good-night ritual. Sometimes Dylan would lie on the floor before climbing into bed, and I would place the bowls around his body and play them. Sometimes he would ask me to put one on his belly to feel its vibrations running through his whole physical being like "sound tickles." His favorite bowls were a little green one, which had an Emerald alchemy, a blue one with an Ocean Indium alchemy, and a little yellow one with a Citrine alchemy. The Citrine vibrated at the throat chakra, the Ocean Indium at the sacral chakra, and the Emerald at the third eye. He liked to play them too, and he loved for me to soothe him to sleep by playing their heavenly sounds. We would say affirmations and pray, each of us adding our own words to create a special alchemical mix of energies that were filled with dreams, visions, and love. It was calming to us both and a most celestial and comforting way to say good night. We loved our little ritual.

Music has brought fulfillment, a multitude of heart-opening connections, and great joy to my life, and it has given me the safety to breathe and feel, to let go, quicken, integrate, and transform. It has taught me to trust that my creativity and authentic expression are good—and that I am enough. Performing has led me to so many unlikely places and venues all around the globe, immersing me in different languages and cultures, infusing in me a broader understanding of what it is about music that unifies us. I hope that as you read this book and we explore sacred vibrations and music as medicine, that you will know you are enough and will experience this oneness that music brings too.

CHAPTER 3

INSPIRATION

Kids4Kids and a Legacy of Love

To whom much is given, much will be required.

— LUKE 12:48, PRINTED ON A MAGNET ON OUR REFRIGERATOR

How wonderful it is that nobody need wait a single moment before starting to improve the world.

**— ANNE FRANK, *THE DIARY OF ANNE FRANK*
(MOTTO OF THE KIDS4KIDS WORLD FOUNDATION)**

When Dylan was almost eight years old, I was moved to leave a legacy of love through music in a country that I came to call home, though I was a foreigner. It was a country colored by both its greatness and by the horror of two world wars. Dylan was half German, and I felt a calling to leave Germany a little better, a little brighter, a little more loving, and somehow to do my part to bring unity and healing through the universal language of music. And so I jumped into the project of creating a children's charity called Kids4Kids World Foundation. All fell into place quickly. The net had appeared. Desmond Tutu's quote inspired me greatly:

Do your little bit of good where you are; it's those little bits of good put together that overwhelm the world.

The Kids4Kids World Foundation was created to nurture children's talents for singing, playing an instrument, composing, writing, dancing, or acting. Our team created an original musical every year with guest directors and choreographers. The children performed a series of five shows, and the money raised supported a music therapy program we created for children in need. This was a win for all.

We taught our international group of children great American standards like George Gershwin's "I Got Rhythm" and Irving Berlin's "Play a Simple Melody," and we included music of different styles and languages. We partnered with an organization called Tabaluga that had been created by German rock star Peter Maffay. Tabaluga had a well-known music therapy program, and every year we raised about $35,000 to purchase instruments for them. We also created a Kids4Kids program and provide music therapy sessions for as many as 20 children per year. They worked with different therapists chosen by our therapy advisor, who was vice president of the European Music Therapy Confederation. It was so touching to watch our young performers grow in confidence, self-expression and professionalism, and to witness the joy they experienced delivering the gifted instruments to Tabaluga, coming together and playing with the children who received our support. I ran Kids4Kids World Foundation alongside my career for almost 10 years, building long-lasting friendships and transforming many lives, until I returned home to live in California in 2013 and the foundation's remaining resources were transferred to Minerva University's campus in Berlin.

Kids4Kids: Improving the World with Music

"Music, in its true expression, knows no social boundaries and can feed the soul when it has been neglected, restoring the human in human beings."

— **Mary Hammond**, *Artistic Advisor, Royal Academy of Music, London*

> "In a world where the focus on materialism and outward appearances continues to grow, it is important our youth have stability leading to independence. Singing and music are steps towards this. Those who have experienced the power of music are less likely to be dependent upon superficialities."
> — **Susanne Klatten**, *Entrepreneur and Sponsor*

Kids4Kids participants competed every year (and always won prizes!) in a national contest supported by the German arts program called *Jugend musiziert* (Youth Making Music). The competition categories changed every year and included flute, piano, musical theater, classical singing, harp, violin, and saxophone, among others. Dylan participated the year he turned 13, and his experience gave me one of the first and most powerful testimonies to the impact crystal singing bowls can have on a person.

He entered in the Musical Theater category and prepared a 15-minute program with five songs, some spoken dialogue, and a dance routine. He sang Duke Ellington's upbeat song "It Don't Mean a Thing," a Gershwin song, a Rodgers and Hart song, and two original pieces we wrote for him. He aced the first round and was chosen to go on to the semifinals. A week before his semifinal audition, he came to me and in an uncertain tone said, "Mom, something's happening. My voice is all over the place." His voice had begun to transition. The larynx of a young man grows larger and thicker as testosterone production increases during puberty, and the sounds that come out of it can be unpredictable. It is a time of vocal vulnerability.

I guided Dylan gently and carefully in some transitional singing exercises. We lowered all the keys of his songs. And we used the crystal singing bowls to keep his voice open and to give him stability and confidence. We had been playing the bowls together for about five years by this time, and they were very familiar instruments for him. He loved playing them. His bowl of choice, as we began working to stabilize his rapidly shifting voice, was the little Citrine bowl, which coincidentally was a G note—the note that supports the opening of the throat chakra! It was a perfect combination of alchemy and note for what Dylan was experiencing. The Citrine alchemy brings confidence, courage, and self-esteem, and it activates personal power.

We practiced gently in small increments daily, never forcefully, and we used the vowel sounds that were most comfortable, like *oh* and *ah*. Dylan sometimes questioned if he really wanted to go ahead and put himself in this musical situation, and I watched the bowls create a container of safety for him. In addition to the Citrine bowl, he used the Ocean Indium alchemy D note and the Emerald alchemy A note, connecting his energy centers of creativity and intuition, which made a sacred interval. An interval is the distance between two notes. These D and A notes, when played consecutively, create the interval of the perfect fifth, a grounded and coherent structure. These notes were exactly what Dylan needed. (For more on intervals, see the glossary.)

We laughed a lot. We yawned often, which helped him continually relax and keep his throat open.

On the actual day of his audition, as he was rehearsing with the pianist, we had to lower the keys yet one more time, one more round of grounding down as he rapidly approached the baritone range.

When it was time for him to step onto the stage, announce himself, and perform his 15-minute program, I watched my son rise to the occasion. He stayed calm and cool. He confidently announced, "My name is Dylan. I'm 13 years old, and as of one week ago, I am no longer a soprano!" The audience chuckled. He sailed into his program and nailed every song as well as his monologue and his choreography. I was amazed and impressed by his presence, humor and ability to perform under the physical pressure of a voice that was shifting gears. He demonstrated such courage, and on that day, we made a memory with the crystal singing bowls that I will always treasure.

Dylan was not the only young person to experience transformation through the bowls in those early days of my explorations with them. One of the older performers in the Kids4Kids productions had a profound healing. She had childhood abuse in her background and struggled at times to allow her voice its full range of expression. Once we began integrating the crystal instruments into her vocal lessons, her voice completely opened and solidified! She also participated in the *Jugend musiziert* contest, and she not only made it to the finals, she also took the overall second prize. The bowls were working their magic, especially for children and young people.

> ### Sound Rx
>
> Melissa, who has become a sound-healing practitioner through the Sacred Science of Sound Crystal Alchemy Trainings I offer, shares this story of an experience she had with a child during the Mother's Circle Sound Bath she runs in Australia:
>
> "One mother came to the sound bath at the last minute, feeling distraught and overwhelmed by her personal life. Her son had autism and anxiety, and he had been bullied at school because he wouldn't play guns with the boys in his class. The mother told how this had affected him and her family. She was so moved after the sound bath that she was speechless. Her eyes, which were wide open, shone like stars. She said she had no words to describe how she felt: it was indescribable bliss. Afterward, her son walked into [my] yoga studio, and he was mesmerized by the bowls and candles. I felt called to play for him.
>
> "His entire face and body transformed. The tension released, and he relaxed. I sang Arielle's song "Listen to the Bowls and You'll Be Happy." (Note: Everyone loves this song, one that Melissa's five-year-old daughter created. You can listen to it on Source, the Sacred Science of Sound app.) He told me he felt happy and calm. Afterward, his mother e-mailed me and told me that he thought I was the nicest lady he had met and that he loved me. He loves the bowls and the healing. His mother said it was like listening to Heaven."
>
> Another of our advanced practitioners in Canada worked regularly with an autistic child who never spoke and who came slowly alive through her regular use of the singing bowls with him. She shared with me her joy at watching his eyes light up every time she came and how he expressed himself in sounds and communicated with her and with his parents in a new way as a result of experiencing the tones of the singing bowls. There was excitement in his eyes. He engaged more in his life. He was awake and present. His parents were overjoyed at his progress. He looked forward every week to his sound bath, and he was changed through the crystal singing bowls.

One of my own experiences working with children was extremely touching. A mother brought her eight-year-old daughter to see me at the Crystal Cadence Sound Healing Studio in Los Angeles, noting that the girl was experiencing anxiety at school. More and more younger children are being labeled with this condition and then prescribed medication. This mother was looking for alternatives. After the first session, the child described herself as feeling peaceful and safe, gently floating above a green meadow, with no more anxious feelings. She was astounded. I also had her play the bowls herself and tone short affirmations she created. Her social interaction and emotional stability improved greatly.

On one level, the bowls provide a scientifically sound mathematics of healing that can be explained through vibrational physics and theory. On another level, what happens with the bowls defies explanation, intellectual understanding, and definition. What is it about sacred crystalline vibration that provides a pathway to the unexplainable, the unfathomable, even? What is it that opens a channel for experiences that touch our soul and take our breath away?

A CHANNEL OPENS

One day after I had passed through the crucible of Dylan's death and was walking the painful path to "normalizing" my life without him in his physical form, I received a call from a friend who was six months pregnant. She had been taken to UCLA Medical Center because there was a problem with the baby. She asked me to come and play the bowls for her. It was a peaceful time together, and the baby and mama received some relief. I drove home feeling glad I could be of assistance. I had almost arrived at the house Dylan and I had shared when I remembered our bike rides and the way Dylan would push me going up the steep hills. He would ride right behind me, encouraging me. I could hear his resonant voice boom, "Come on, Mom, you can do it!" And right then and there, I lost it! I missed him so badly in that moment. I started yelling and shaking my hand at the sky. "Why can't you

Inspiration

be here? Everybody loved you so much! You are so missed. Dylan! Why can't you be here?"

In that instant of my passionate outburst, my car stereo turned itself on and started playing Louis Armstrong's rendition of "What a Wonderful World," one of Dylan's favorite songs. Every hair on my body stood on end, and I stared at the car's audio screen in disbelief. What was more unbelievable was that the name *DYLAN* was printed on the screen in capital letters. I pulled the car over as Louis crooned, "And I think to myself, what a wonderful world. . . "

I took a photo of the screen to show to my family. I could not believe it! Dylan was talking to me through music! "Unknown genre," the stereo said. (Oh, yes.) "Unknown Album, Unknown Artist. Back, Home, Favorite." The number 77 was there, a double seven being a sacred number representing a grounded spirituality and mystery. The numbers 54 and 63 were also on the screen! The numbers five plus four, and six plus three, both equal nine: the day of his birth. He was bridging a world I could not see.

Dylan communicating with me through music, as shown on this screenshot I took of the audio screen in my car as it played Louis Armstrong singing "What a Wonderful World." (Photo by Jeralyn Glass)

Dylan communicating with me through vibration had become a regular occurrence since the extraordinary shooting star in Los Angeles and his exclamation that he was "Home and with God." We had conversations daily. Sometimes when I would lose the bigger-picture perspective and drop into sadness or not find any motivation to go on, I'd receive an intense message accompanied by physical signs of his presence. A few months after Dylan communicated with me through the car radio, he sent a remarkable communication, one that gave me a new perspective on this life and our eternal life.

DYLAN'S VISITATION

I was in my early period of exploration and daily practice with the bowls as healing instruments when the news came that the owners of the home where Dylan and I had been living wanted to move back into it. I was feeling tender and vulnerable and wanted to hold on to every piece I could of my physical life with Dylan; I had come to think of that rental house as our home. It was a strange comfort to walk up and down the stairs Dylan had walked, hang out in the living room where we had laughed and shared so much, and to sit in his room, feeling him, sensing him, smelling him, remembering details of our life together. I had not yet had the courage to go fully through his things, and almost everything remained exactly as he had left it. I had given some of his clothes and athletic equipment to a charity organization that supported young athletes and some to his close friends, many of whom had asked me if they could have this shirt, that hoodie, or that book. His community was grieving him too.

One evening, after spending reflective time in his room, I was very restless. The move was hard for me. I was struggling to have to leave the house we had lived in together. Back in my own bedroom, I fell deeply asleep and had intense dreams. In one, Dylan was with me, larger than life, and as real as when we had been together. He appeared first right above my bed and seemed to reach down and lift me up and hold me. It was so real. He was warm; I could feel his arms wrap around me in his gorgeous bear hug, hear his breath, smell the cologne that he loved so much. He was talking to me. "Mom, it's all okay. I'm fine. We've got this. I'm safe. You're safe. Please don't worry." I fell back asleep held in his comforting presence.

When I woke up the next day, I jumped out of bed and ran downstairs to his room to say good morning! He was back! My heart was pounding with excitement. His death had been a crazy dream. I had no time to think—only to go hug him with incredible joy! As I entered his room, I looked around in surprise. Everything was exactly as I had left it; there was no sign of Dylan. I shook my head in despair and dropped to his bed, recognizing the absolute absurdity of what I had just done. I could not believe my naiveté. I crawled up the stairs and sat myself down in his favorite chair.

As I let my head hang, my eye caught something in my nightgown: buried between the threads of the flannel material was a white feather! A single, white feather as if woven into the fabric. In spiritual circles a feather is known as the sign of an angel. It startled me. I stared at that feather; then I grabbed my cell phone and took a photo of it. It was embedded under one thread. I am highly allergic and have been since childhood, and due to my allergies, I do not have anything made with down feathers in my house—no pillows, no quilts. My heart lit up in wonder. Dylan had sent me a profound sign. What I had experienced was a visitation, although I did not know it had a name. He had come to me. Everything I had "dreamed" had occurred, except that he could not return to his human body. This experience was phenomenal. It gave me courage and strength. The veil between this life and the next is thin.

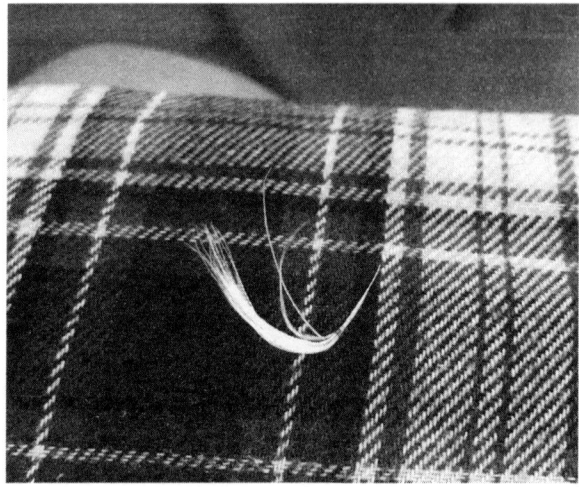

This sign of Dylan's "visitation," a single white feather tucked under a thread of my flannel nightgown, gave me courage and strength. (Photo by Jeralyn Glass)

A BRIDGE FROM THE SEEN TO THE UNSEEN

I developed more resilience through Dylan's communications and had a treasure chest of hope stored up when the grief and longing would hit again. What I did not realize was that I was increasing my ability to stay elevated in my vibrational frequency. Playing the singing bowls every day and integrating embodiment exercises after Dylan's death helped immensely to release what was not serving the highest good. Those relationships that were not a resonant match gently faded from my life; other existing friendships deepened, and new friendships blossomed. Sound was fervently composing a symphony that was changing my life.

We must be willing to trust in the perfection of our lives as they roll out, often so different from what we had hoped, from what we had dreamed, from what we had intended.

Our task is to trust that our ultimate good is unfolding, no matter how it seems. I would have gladly given my life if my boy could have lived. In a heartbeat. Yet, that was not possible. Our destiny was a different one. Dylan was heading out first, and Mom would remain. Together, we bridge Heaven and Earth. Sublime, indescribable music. Crystalline sounds beyond this earthly plane. Celestial tones. The authentic expression of two hearts broken wide open. And two daring souls who said yes to this destiny. George Bernard Shaw wrote that the true joy in life is being used for a purpose recognized by yourself as a mighty one. I could only view the loss of Dylan from this physical plane as a sacrifice; somehow, we were both being used for a purpose much bigger than ourselves. And if I jumped, if I said yes, would the net be there this time to catch me? I had no choice. I had to accept. I had to learn to have faith.

My oldest sister reminds me that Dylan lived more in his 19 years than many people live in their lifetime. Looking back on our time together, it's clear my sister is right. His life was a brilliant kaleidoscope of experiences and emotions, and it has turned out to be, in truth, a vast and unquenchable existence that has found expression through sound vibration that takes us beyond and brings us back to Earth again. Connecting the unseen to the seen. My invitation to you is that if you are struggling with something that is unbearably painful or insurmountably difficult, please know that resolution is possible. Mysteries occur all around us, confirming we are not alone. Peace can find you in the most unexpected ways. Sacred vibrations can guide you home. Some may call that mystical, and maybe it is. But it's also a reality, one that can transform our lives and our potential more than we ever thought possible.

CHAPTER 4

EXPANSION

A Bigger Purpose

The pain I feel now is the happiness I had before.

— **C. S. LEWIS, *A GRIEF OBSERVED***

The first call came in at 2:22 with unfathomable and unacceptable news. There had been an accident, the woman on the phone told me, and Dylan was in the hospital.

Then the second call. Dylan had not made it. I was in complete shock, a barely functioning shadow of myself, ashen and numb.

Then a third call came, and the stranger on the other end asked me what I would like to do with Dylan's organs. That was it. I had managed to keep myself together until that moment, and then I shattered into pieces. The worst part of receiving that call was that I could not ignore the reality of Dylan's death or that parts of his body could help others. Only a few hours after receiving the initial confirmation of his death, I was forced to face the truth and make a decision. They needed an answer. Now. I could not even comprehend that he was gone, let alone decide what to do with his human body.

Did I need him to remain whole? Would I donate his organs and body parts? Would I cremate him? Bury him? I had never asked myself these questions or spoken with Dylan about his wishes. I wanted to stop the clock. Bring my boy back to life. I wanted to disappear. To be swallowed up by the earth. Scream. I wanted to stop the pain immediately, yet nothing I did could stop the anguish—nothing I could ever do would bring Dylan back. I was overwhelmed, yet I understood that whatever I chose to do to relieve that pain would only be a temporary Band-Aid on top of the helplessness, the emptiness, and the despair. There is no pill and nothing to comfort you or ease the grief when you unexpectedly lose someone you love. Or when you are thrown into the arms of anguish through whatever unexpected situation confronts you. Nothing could ease the indescribable hopelessness I was feeling.

In the days following Dylan's death, my choices began to reveal themselves little by little: I needed to go inside and listen deeply, to give myself the space to sit in stillness, to ground myself as best I could to stabilize the wailing and the wobbling. There is no one way to process pain and sadness, no single way to button it up that fits everyone. Indeed, pain, both emotional and physical, is a very individual and personal journey for each of us, and the discovery of the pure quartz crystal alchemy sound as a balm of healing brought the light and frequency of love into the darkness and answered my cry of helplessness.

Although the idea of a bigger plan, a Divine destiny, was not completely new to me, when I was confronted with the pain of losing my son, the concept of me as creator, as having agreed to this loss, was incomprehensible. Yet whether I liked it or not, by grounding myself in the great pain, by not running away and numbing out, by landing in and embracing the fear, and with a net held by sacred vibrations, the deeper learning process had begun, and healing was happening.

There were many physical signs confirming this path. Dylan died on a Friday, and Easter Sunday fell the week following his death. On that day, I found myself walking down to the beach and saw a gathering of people and a wooden cross. A church service was being held. I sat down on the sand and looked up at the bright, blue sky. A rainbow in the form of an angel appeared for an instant. Trembling, I reached for my phone and took a few pictures of what I saw. And then the rainbow angel disappeared. This was a message of hope. I felt my son comforting me, although I could not possibly understand the immensity of it yet.

This rainbow angel appeared on Easter Sunday the week after Dylan's death and was a message of hope. (Photo by Jeralyn Glass)

We never know what lies ahead and it is in those moments of deepest devastation that we learn who we are. Instinctively I knew I needed to prepare an earthly closing for me and Dylan. But how? What would that look like? I had no reference point. But I knew I had to anoint his body and this sacred ritual awakened in me a timeless memory of having done this before, somewhere in an existence hidden by the mists of time. It quickened me into the understanding that the veil between this life and the next is thin. As I gazed at his body, I realized that Dylan was no longer there, but he was not gone. We are eternal. I tenderly washed his feet and then his hands, gently rubbing them with a special oil we had bought together on the Big Island of Hawaii. I was not frightened of death as I feared I would be. I was able to embrace it, to cherish it, to honor the time Dylan and I had been gifted. Death was a natural part of life. It was soft and vulnerable. I remembered the poignancy of devotional love while playing Mary Magdalene as a young singer, and the indescribable sweetness it embodied in me. I recalled standing before Michelangelo's Pieta in Italy—the famous marble statue of Mother Mary holding Jesus's body after his death. I knew that the exquisite energy of the Divine Feminine Presence was tenderly cradling me. Her vibration was palpable and brought great comfort. To accept Dylan's death being held this way in Grace was the most precious, intimate, and unexpected experience I've ever known.

I said goodbye to my son with reverence, reading him endearing goodbye letters, and sharing with him little gifts from family and friends, including a bunch of wild sage we had picked and bound together when he was 12. "Be Courageous, Jeralyn," I heard, a whisper from above. I closed the thick metal door for his cremation. A new relationship with him anchored in that defining moment.

After the mortuary visit, I returned to the hotel in silence with the small group that had accompanied me.

That evening, I asked Peter to take a walk with me under the night sky. As we walked the woods far away from the hotel, I held his hand and started making loud and jagged sounds of grief, groaning, and yelling, tones that expressed despair I had not yet released. I was held by the velvet softness of the night sky and the depth of my dear friend's compassion. We both looked up to the heavens, just as we had done on March 27, the night Dylan died. This time, we saw a dazzling array of shooting stars, a stunning display unlike anything I have ever seen in my life: a celebration in the night sky, among the stars Dylan and I both loved so much.

IMMERSED IN EMOTION

What things bring us comfort when we are emotionally raw, and how do we access them when we are in pain? I had to learn how to answer these questions. And I had to do it in the way that felt right for me. There is no one way and no prescription to process grief.

It's interesting to note that three different doctors advised me to go on antidepressant medication after Dylan's death. I knew how wrong this choice would be for me as a means of dealing with the loss of my son. I knew it would numb me out and I might never come back to myself. And I knew the real outcome I desired would not be possible: medication would not bring Dylan back. As scary and lonely as it was, I knew I had to feel the loss and all the feelings that went with it: regret, remorse, shame, guilt, anger, fear, despair, shock, disbelief. I knew I had to be strong enough to feel it all, or I could not heal, and I would never be able to reclaim my life and full energetic expression. My antidepressant and pain-relief medication of choice was crystalline sound vibration, and it did everything I asked it to do—and more—with no side effects.

The reality is that pain—emotional, physical, mental, and spiritual—is all around us. And it has enormous costs. We have become isolated, separated, fearful, and confused, and we are seeking more ways to be in charge of our health and well-being.

THE SEARCH FOR SOLUTIONS

The management of pain has given rise over the past few decades to a massive search for solutions. Painkillers have long been the first line of defense for pain management, but they eventually stop working, and they come with some significant downsides that are acknowledged but generally ignored. We are also more aware than ever of adverse drug reactions and "bonus" problems that invite prescriptions for additional medications. What happens when conventional medicine fails to relieve the pain people are in?

Fortunately, some wonderful alternatives have become more mainstream. Sound therapy and energy medicine are now being recognized as effective pain-relief agents. The crystal singing bowls' healing tones have emerged as a strong "peoples' choice." Tuning forks, gongs, Tibetan singing bowls, drums, mantra, didgeridoo, harp, and singing are also all being successfully implemented. Many are also using acupuncture, chiropractic, osteopathy eye-movement desensitization and reprocessing therapy (EMDR) and ultrasound instruments to alleviate trauma and pain. Breathwork is essential to release held emotions. Some people are choosing to use plant medicine and the microdosing of psychedelics and psilocybin (the chemical found in "magic" mushrooms) as well. A gentle reminder: self-love and self-care are important ingredients to add to the recipe of how we generate healing.

The bioenergetic healing techniques that Dr. Sue Morter teaches represent another powerful solution. These are composed of nonforceful, energy-balancing practices that can be either hands-on or remote to help reestablish the full healing potential of the body and bring it back to its natural state of health and wholeness. Dr. Sue's process guides the individual into higher states of consciousness through the language of the body, producing a perfecting effect on every level of life. Her state-of-the-art healing has been an important part of my journey through grief. It has helped me become aware of my subconscious beliefs and patterns, to release and clear them, and to build healthy neurocircuitry. Healing my grief happened as a by-product of awakening to and recognizing my true nature, of quickening into the being I am here to be. As Dr. Sue Morter so beautifully exemplified, healing happens in our lives as our true magnificence opens and expresses itself. This is at the core of our transformation.

SOUND HEALING GOES MAINSTREAM

I noted in an article I wrote for *Global Health Resources* a few years ago that sound healing has become a mainstream buzzword, and now more people than ever are experiencing the benefits of frequency and vibration in their search for health and wellness. Many people are seeking more ways to practice mindfulness and find well-being, inner peace, relaxation, and regeneration as we struggle to reinvent life in a post-pandemic world. Stress has increased with the fast pace of living and the extreme pressures that young people in particular are facing as a result of negative social-media influences.

The popularity of meditation apps such as *Headspace, Calm, Healthy Minds* and *Insight Timer,* (with hundreds of millions of downloads) confirms the heightened interest in new approaches to health and wellness. The Sacred Science of Sound released the app Source in the spring of 2024. The app idea began in 2020 when Dylan nudged me to create a collaborative platform with scientists and musicians. This is a one-of-a-kind tool you can hold in the palm of your hand, and it brings you cutting-edge science, energy medicine, meditation, yoga, tuna forks, sound baths, and the healing power of music. It's geared for all ages, and even includes meditations for children.

Out of my grief came a desire to help others, easing my attention away from my own. I reached out to work with a nonprofit dedicated to helping cancer patients and their loved ones, and I created a program for them in 2015. The following year I began playing for hospice patients and their families, and veterans with post-traumatic stress disorder (PTSD). The feedback was humbling.

Nancy Lomibao, program director at the Cancer Support Community in Redondo Beach, California, noted that she has frequently heard from the organization's participants about how one or another of their classes has impacted them, "but nothing as dynamic and passionate as the feedback I hear about Professor Jeralyn Glass's Crystal Alchemy [sound] bowls class. Our participants tell me it's 'magical' and 'transformative.' They rave about how they are pain-free for that hour, that they can finally concentrate afterward, and that all their anxiety and stress floats away. Our participants absolutely love this class for the calming reprieve and relief it brings to cancer."

What *are* these state-of-the art sonic instruments, the alchemy singing bowls, that I use for the sound-healing classes? Each bowl is formed from high-grade, pure quartz and then infused with semiprecious gems, minerals, and metals. Pure quartz is known to be an amplifier, a transmitter, and a receiver, and the quartz crystal bowls amplify, store, focus, transfer, and—most importantly—transmute energy.

Quartz is the ultimate vehicle of communication, and its use is widespread in our daily lives. It is an essential component in computers, cell phones, automotive electronics, radios, and microphones. We know the human body is crystalline in structure, and the sound and light emitted by quartz crystals provide deeply healing effects on our organs, muscles, tissues, and cells, as the work of my esteemed colleague John Stuart Reid has shown.[1] This structural similarity makes it effortless to receive the quartz singing bowl vibrations. Imagine that, like the singing bowls, we too are vibrating crystalline vessels, each with our own unique frequency signature. It is ours to discover and to express that personal vibrational signature. Every tone and vibration emanating from the bowls provides a benefit through frequency and sound. Every musical note corresponds to an energy center in our body, aligning and balancing our human system.

Quantum physics has revealed that atoms are made of vortices of energy, each radiating its own energetic signature. From the level of an atom to the farthest reaches of the galaxy, everything in creation has a frequency. Every thought and feeling, every organ, and every part of our body has a vibration, even every illness. When the vibration is in harmony with the rest of the body, we experience health. When it lacks harmony, we experience dis-ease.

In crystal sound therapy, the pure, high-frequency sounds resonate and entrain with our physical, emotional, and energetic bodies, bringing the possibility of clearing, cleansing, and balancing at the cellular level.

Expansion

Every class I teach for cancer patients begins with a playful introduction exercise that invites them to use their voices in creative expression. We set a group and a personal intention, and then a 64 Hz weighted tuning fork is placed on the soles of each patient's feet, helping them to feel the physical sensation of sound through the vibrating tuning fork. For those with neuropathy, we find a place on their body where they can feel the sound; sometimes it is on the calf or thigh, sometimes on the hand. After they receive sound vibration through the weighted tuning fork, they lie down on yoga mats and are then bathed in crystalline harmonics accompanied by a guided meditation. The primary goal is to reduce stress and pain through deep relaxation, which in turn develops self-awareness, promotes creativity, improves learning, increases strength, and clarifies personal values. I also like to use beat frequencies (the third tone created by the brain when we listen to two tones at different frequencies). I have found this shimmering sound effective in supporting emotional balance and mental clarity, which can be challenging for someone with cancer to sustain due to the anxiety and fear often caused by their diagnosis.

Sound Rx

Here is some of the feedback I've received from my sound-bath work with cancer patients:

I find the most profound benefit is my ability to relax. Sometimes, I've gone into the session quite stressed and upset. The music and the atmosphere have 'overpowered' me, even when I've fought against them and kept obsessing about my problems. I've felt myself drawn in and unable to keep obsessing. By the end of the session, I'm relaxed, refreshed, and able to come up with ideas about how to positively deal with my situation.
— Nancy

I am a cancer survivor of a few years, living a full life. I had the privilege and blessing to experience the crystal bowls in a group setting in 2017. It was an amazing deep healing meditation which made me want to repeat it. I experienced numerous benefits. Each time it's a new experience and so personal, even in the group setting. The combination of the sound with Jeralyn's beautiful voice shifts my focus. I'm living fully in the moment with confidence and joy, sharing with others. My memory, which was severely impacted by chemotherapy, has greatly improved.

—*Diane*

Ten years after a cancer diagnosis and the concomitant removal of various body parts, I could no longer face disease alone. Among dozens of activities offered at the Cancer Support Community, crystal bowls did not fit my academic background. Better a lecture on how to cure a swollen prostate and a depressed brain. Many of my new cancer warrior friends praised their vivid experiences—physical, aural, and emotional—calmed nerves, ameliorated pain and discomfort, with sound vibrations that penetrated body, bones, and tissue.

The first session took me beyond my academic sense of healing. Two series of crystal sound vibrations surpassed what I had thought my body capable of. The first filled a horizontal rectangular area of my upper chest about 12 inches by 4 inches and about a half inch beneath the skin of my chest, lasting more than one minute. The second one went far beyond the first in vibration, length, intensity, and wonderment. The inside of my head became a vibrating crystal bowl first from ear to ear, then inside the cranium, circling and banging and crashing into my inner cranial bowl, bouncing playfully, back and forth, round and round, blissfully for nearly three minutes. After a year, I can still relive some of the vibration and enjoy it again. The crystal bowl therapy eased the psychological pain of my cancer, creating new cognitive experiences and new corporal pleasures.

—*Kenny*

Since my diagnosis, Jeralyn's crystal bowl class has definitely changed me. It is not only how gifted she is in music; it is her smile, her heart, and her soul that come through and shine in that room and the peace she brings with her crystal sounds. I don't know exactly how, but it gives me an overall soothing in my body, mind, and soul. I have gone in with anxiety, restless legs, and aches and pains, and I am always able to relax, enjoy, and feel the "floating" experience beyond space and time—just *bliss*! It is like a whole-body massage and a spiritual renewing!

—*Cony*

When I am creating a sound bath experience with the crystal bowls, I invite listeners to pay attention to the areas of their body and the feelings or memories that may become activated. Emotional and physical traumas can be recorded in the fascia system of connective tissue throughout our bodies, and sound may help those energies to be felt and released. The harmonics of the bowls may also activate increased perceptions of colors and geometrical shapes. People are often surprised by where these exquisite crystalline sounds transport them: to other places, other times, often evoking long-lost memories or connections with loved ones who have passed. I remember when I first heard the bowls more than 17 years ago. Their vibrations were pristine and pure, and they sang a song that was so familiar, something I knew; they produced frequencies that my soul longed for. When we use sound with intention, when it is in combination with integrity, presence, courage, and openness, we transmute and transform. We quicken. We uplevel. Crystalline sound brings a recalibration and an updating of our human system.

DYLAN'S JOURNEY

When Dylan was in his junior year of high school, the six sports-related concussions he had experienced from age seven onward began to trigger intense migraines that interfered with his ability to concentrate. I watched his internal compass waver. He had always been an excellent student, and this was the beginning of a high-pressure time when grades, tests, and writing assignments mattered for college applications. The symptoms increased incrementally, and they finally worsened to the point where he asked for help.

We had tried a few alternative remedies such as biofeedback, personal coaching, meditation, and psychotherapy and were then guided to a leading brain doctor who specialized in working with athletes with concussions and high-impact injuries. He was known to be a great proponent of natural remedies and nonpharma solutions, which is why we chose his clinic. He did several tests and really did not find any strong reason for the migraines or see any damage from the concussions. He found a slight irregularity in the prefrontal cortex and advised Dylan to begin an antidepressant in combination with an anticonvulsant drug that works on the chemical messengers in the brain and nerves and is seen as a possible aid in reducing headaches.

I was totally surprised by this recommendation: I had not expected medication, particularly since one of the side effects of the prescribed medicines was suicidal ideation. I knew my son well. He was intelligent, athletic, physically strong, and very sensitive. He was struggling with sleep. The migraines were excruciating. The doctor informed me that since Dylan was 18 years old, I had no choice in the matter of medication. Dylan was an adult, and under California state law, it was his right to medicate if he so chose.

When we left Germany, where Dylan had been born and raised, and he began high school in the United States, we were both surprised by how many of Dylan's fellow students were on medication. It was uncommon in Germany for physicians to prescribe medications to children, teenagers, or young adults. There had to be a very strong positive indication for a doctor to do so, and German doctors would first offer many alternative mental health treatments, including music therapy, vibrational medicine, homeopathy, and osteopathic medicine.

While Dylan attended high school in California, I taught in the performing arts school of a private university and was astonished and similarly surprised by how many of my students were on medication, sometimes several different kinds. It seemed the norm in the United States to medicate young people, while in Germany, it was not.

Dylan hoped medication would help him, but it couldn't. His fate was directing him to a different path. How I wish I could have had the chance to work once again with my son and the singing bowls as we did in his early childhood and when he was 13. Could I have possibly changed his destiny? There is no answer to this question. It was not meant to be. There was a bigger plan in place.

EMBRACING THE PAIN

Until I faced the death of my child, I had known my identity as a wife, a mom, a friend, a creator, a professor, a singer, a sister, and a daughter. But I had yet to discover that the truth of who we are reveals when we have no choice but to *become*. I had to become fearless in facing a world without Dylan, and I had to be able to anchor myself in the strength of my heart—a heart that had learned it could bear intense pain and did not have to run from it. I had been terrified to experience that reality. I knew that medication was not the answer for me, yet I recognize that the road to mental health and emotional balance is a very personal one.

The courage that began to anchor in me in the days and months, and now years, that followed Dylan's passing has been unexpected. Returning home after Dylan's cremation, I could barely face my daily life without him. I completed the semester teaching at the university, knowing I was no longer the woman I had once been. I was in two-hour grief therapy sessions three times a week, struggling to process Dylan's death. I was also trying alternative ways of healing. The first few years were filled with ups and downs. There were moments where Dylan was communicating with me through signs and frequency, but while I was paralyzed by grief, I could not easily access his energy. The deeper I landed in my grief and my human story, the more distant he became. I knew the dance was to stay in a higher vibration so that I could connect more easily with him. Sometimes, it was challenging.

One day as I was walking with my dear friend Peter down to the beach, talking about Dylan and his childhood in Germany, an acorn was propelled out of a tree as if someone had thrown it down upon us with vehemence to get our attention. The acorn landed directly next to my right baby toe. Peter and I looked up at the tree above us. No squirrel. No bird. No person! We laughed. Dylan was communicating.

"Mom, there is so much work to be done! Don't suffer! I'm here! I'm always with you!"

My tears flowed when I heard his words, but through them also came the joy of knowing my son's dynamic presence was still active in my life. He was reaching for me. In time our closeness and connection through sound vibration solidified, and the celestial tones of the bowls helped tune me to a frequency that would facilitate our communication. I am still in awe of what they, other sound-healing instruments, sacred intentional music, and our own voices can make possible in our lives.

As life unfolds for each of us, we come to understand the importance of accepting that our lives are perfect as they are. There is a bigger purpose, and as we learn to face our personal dangers and our challenges, unafraid to feel them, breathe, and process the pain, we expand and strengthen our hearts. Every moment of our lives offers us a choice. We can choose a higher vibrational frequency that includes acceptance, not judgment. A frequency that helps us to ground, to forgive, and to choose thoughts and actions that support emotional stability and inner peace. Music medicine's secret ingredient is Love, and Love is the vibration that transforms.

CHAPTER 5

CONNECTION

Sound, an Ancient Medicine

Pythagoras was likewise of the opinion that music contributed greatly to health, if it was used in an appropriate manner [and that]... at another time they used music in the place of medicine.

— IAMBLICHUS, 4TH-CENTURY PHILOSOPHER, FROM *LIFE OF PYTHAGORAS*

Rhythm and harmony find their way into the inward places of the soul.

— PLATO

In the 1980s in Silicon Valley, fused-quartz crucibles were being used as containers in which to grow crystals at high temperatures for the semiconductor and computer industries. During the production process, some of the bowls cracked, chipped, or broke, and some remained intact. It was discovered that these pure quartz crucibles could be repurposed as musical instruments, and in the right hands, they could make heavenly sounds. Around the year 2000, the first generation of alchemy quartz singing bowls was created by blending pure quartz with precious gemstones, metals, and other organic substances. These modern instruments have quickly become one of the sound-healing instruments of choice.

Those who hear their resonant song respond deeply and intuitively, for their harmonics are nothing less than magnificent. Each singing bowl is a one-of-a-kind instrument. Their unique overtones and frequencies emanate from the alchemical powers of the quartz combined with additional special elements like Rose Quartz, Charcoal, or Gold. When the bowls are played, the vibrations of these "alchemies" are amplified and promote awakening, activation, bliss, discovery, joy, clarity, inner peace, and the transmutation and transformation of long-buried, stagnant, and blocked energies.

The bowls themselves are sonic pieces of art that give forth a beauty and exquisiteness of tone that is palpable and powerful. The size of the bowl and the thickness of its walls affect the notes and the overtones with which it resonates. The singing bowls amplify pure vibrations that can be experienced as grounding, centering, or accelerating, and they are accentuated and felt in different parts of the body. As we understand the basic principles of bio-energetics and the science behind sound vibration and frequency, our connection to sound as an ancient medicine becomes more real. The memory of unresolved emotions stemming from trauma and trans-generational patterns gets stored and buried in our cells, which can create a range of symptoms and illnesses. Crystalline sound may help us reveal these emotions and patterns so we can transform them. The opportunity to lovingly reconnect with our wounded parts reawakens the healer within and restores the memory of wholeness into the system.

THE HISTORY OF SOUND HEALING

We know now that everything is energy and sound, including us, and that sound can affect matter, thus creating physical changes. New research and discoveries from Stanford University are opening phenomenal possibilities showing that "acoustics can create new heart tissue. Frequency and amplitude put cells in motion, guiding them to a new position and holding them in place. Sound can create new tissue to replace parts of a damaged heart."[1]

Every atom, molecule, cell, gland, and organ of the human body has a vibration and emits sound, and like any excellent orchestra, the whole must be harmonically tuned to reflect good health. With the loss of my son, I also lost my "tuning." I found myself on a journey to recalibrate and find my new tuning, observing and learning from distinct cultures. I traveled to other countries and began to explore how sound vibration was used as a medicine in different traditions. Discovering and experiencing music in this way brought an understanding of the depth and possibilities inherent in sound from ancient times to the present.

I learned that the documented use of therapeutic sound vibration stretches back to the beginning of recorded history. When I traveled to Australia in 2017 to present at the first Crystal Bowl World Sound Symposium, I learned from the native elders that as long as 60,000 years ago, Aboriginal tribes used vibrations from their indigenous instrument, the didgeridoo or *yidaki*, to heal tissues and mend broken bones.

In India and Bhutan, surrounded by the breathtaking landscapes of the Himalayas, I experienced powerful mantras in Sanskrit that I had never heard before. These sacred sounds have been used for thousands of years to balance the chakras and awaken and elevate consciousness. The powerful vibrations of the healing mantras accompanied me as I hiked up to 14,000 feet in the Himalayas and shared Dylan's ashes with the mountains he loved so well. Toning mantra brought stillness and acceptance to my heart. They anchored a new sense of compassion within me. *Om mani padme hum* and *Ra ma da sa sa se so hung* were chanted as we ascended the mountain and during the intimate celebration of life. As the ceremony finished, tears were streaming down my face. I had gently closed my eyes to be in oneness with Dylan and the majesty of these sacred mountains. As I opened my eyes to share my gratitude and say farewell, I was astounded to see that right next to Dylan's ashes lay a leaf in the shape of a heart.

My travels to Peru revealed the vastness of the ancient knowledge that shamans and medicine people had about music, dancing, drumming, storytelling, and using their native instruments. With their drums, rattles, and bells, they chased away negative energies. They also played the flute to invoke the protection of sacred animals and bring health and harmony. I experienced the fire ceremony that symbolizes the burning away of all that no longer serves. And most importantly, I learned that singing to the earth was a blessing necessary to support our human journey. Gracias, Pachamama.

In Chile I meditated, played my bowls, and toned in front of the Moai giants of Easter Island. Being face-to-face with those fantastic megalithic statues awoke within me ancient memories of the infinite possibilities with sound. There was an energy of the mysterious, the universal, the immutable, and the ageless. To my surprise, I rediscovered the sound of my own laughter, which I had completely forgotten. Dylan was present, laughing his big belly laugh, reminding me that our time here is short and we are just a tiny speck in the great Cosmos. I felt so held and so light at the same time. These sacred vibrations had returned me to a pure state of childlike joy and innocence; I marveled as my laughter flowed freely in a way it had not done since his passing.

In Tonga, an archipelago in the southern Pacific Ocean, I swam with humpback whales, connecting to vibrational communication beyond any known language. I wanted to get close to these legendary creatures and feel their sounds. Being underwater with those gentle giants was indescribable. I swam with a mama and her baby one day at sunrise, and surprisingly they were there again the next morning. Seeing their interaction two days in a row brought a reflection of hope. Hearing their bygone mystical sounds and interacting with them in the warm ocean, feeling Dylan by my side, helped me understood that a vibrational connection between a mother and child is eternal. Inspired, I later created *The Crystalline Sounds of the Earth*, an album of 432 Hz tuned crystal bowl music that included a track infused with whale songs recorded in those magical waters.

ANOTHER ANCIENT CULTURE: TONING IN THE KING'S CHAMBER

I immersed myself in the awe-inspiring culture of Egypt, experiencing the energies of the Great Pyramid—actually lying inside the sarcophagus and toning in the King's Chamber—exploring the temple of Dendara and the Hathors and imagining life in the ancient Egyptian sound-healing chambers. One evening, as I stood on the balcony of my hotel room with Dylan's photo in my breast pocket, I was filled with an immense and palpable loving energy. I was gazing out at the pyramids when this incredible energy vortex revealed itself. I saw orbs of light, round as the full moon, through my camera lens. Some of the orbs had intricate details. I had never seen anything like that before. Thinking about it now, to me they resembled the geometries of sacred sound, that we know from the cymascope images.

For the first time, in Egypt I saw orbs of light, vibrating at a loving and awe-inspiring frequency, surrounding the pyramids. (Photo by Jeralyn Glass)

Pythagoras, the Greek philosopher and mathematician, is said to have studied in Egypt around 535 B.C. He believed that both mathematics and music had the gift to purify the body and soul and that music was indeed that which connected us to the

Divine. His writings on frequency and music's ability to transform illness has been highly influential in moving music into clinical settings, as well as to the development of the modern world of sound in therapeutic settings.

EVERY CULTURE UTILIZED SOUND

My dear friend Jonathan Goldman, whose research I shared in Chapter 1 on the healing potential of humming, conducted with his wife, Andi, notes that the therapeutic use of sound as a tool for healing and transformation has been found in almost every culture and every tradition on our planet. And with great gratitude, I thank him for his enormous knowledge of how sound healing works, as well as for the following contribution:

Many ancient traditions affirm that the universe was created through the power of sound—for example, in the book of Saint John, in the New Testament of the Christian Bible, it is written: "In the beginning was the Word, the Word was with God, and the Word was God." This is confirmed in many other religious texts as well:

- The Vedas of the Hindu tradition note that "in the beginning was Brahman, with whom was the Word. And the Word is Brahman."

- The ancient Egyptians believed that the god Thoth created the world by his voice alone.

- The Hopi nation tells the story of Spider Woman, who sang the Song of Creation over the inanimate forms on Earth, bringing them to life.

- The Aztec legends tell the story of the creator, who sang the world into creation.

- In Polynesia, there were originally three great gods who created the world through the sound of blowing the sacred conch.

- Many African legends from different tribes tell of the creation of the world through sound.

- From the sacred Mayan text of the Popul Vuh, the first humans were given life solely through the power of the word.

In addition to the various sacred texts describing sound as the major force of manifestation, the ancient mystery schools that existed thousands of years ago in Rome, Athens, Egypt, India, China, and Tibet had vast knowledge of the power of sound to heal. The writings that have survived from those times indicate that in those schools, the use of sound as a therapeutic tool was a highly developed spiritual science based upon an understanding that vibration was the fundamental creative force of the universe.

In India, there is a saying: "Nada Brahman"—the world is sound! The words of the ancients are now echoing those of our top scientists, such as quantum physicist Michio Kaku, who through his statement "the universe is a symphony of strings, and the 'Mind of God' is cosmic music" is telling us that everything is music. Modern physicists, in fact, tell us that this dimension and other dimensions are composed of tiny strings that vibrate at different rates.

Let's take this a step further: if everything is music, and everything in the universe has vibrational rates, then our bodies do too. Every organ, bone, and tissue vibrates to its own specific, healthy, and natural resonant frequency. When we are in a state of health, we are like an incredible organic orchestra that is playing the "Symphony of the Self." In fact, when we are feeling good, we say we're in "sound health."

But what happens if the second violinist in the orchestra of our body loses their sheet music and begins to play out of tune and out of harmony? Soon the entire string section sounds off-key and the whole orchestra begins to suffer. This is akin to what happens if a part of our body loses its natural resonance—it begins to vibrate out of ease and out of harmony. We say it is "dis-eased."

The solution to this difficulty is something most holistic health practices focus on: restoring the sheet music to that string player and somehow projecting the correct resonant frequency to the part of the body that is vibrating out of tune. This idea of restoring balance is the basis of chiropractic, acupuncture, chromatherapy (light therapy), aromatherapy, nutritional therapy, and many other practices. And it is what happens in sound therapy.

Jonathan's orchestra analogy wonderfully expresses my journey of re-tuning myself. This is available to all of us. Whether we name it "sound therapy," "sound healing," or "music medicine," it is the balm humanity needs to reactivate in our daily lives, and the crystal singing bowls are a powerful entry to that resonant tuning of health and well-being.

Sound Rx

One person who has used the crystal sound-healing bowls with great impact is one of my former students, Dr. Sole Carbone. She is a doctor in traditional Chinese medicine, psychotherapist, and polymath committed to the principles of integrative wellness and the unification of science and spirituality. She is a specialist in amplified states of consciousness without the use of chemicals—especially through the transformative power of breath, evocative sound, and crystal bowls. She is particularly focused on the role that our inner frequencies have in the healing process and the relevance of coherently tuning the mind, body, emotions, and spirit.

"After almost 20 years of using sound in a therapeutic setting, I can see that, besides its recognized beneficial properties, sound seems to have the capacity to function as a hinge," she says. "That is, it can behave as a device to access deep layers of the subconscious mind and dissolve the resistance of the ego-self. When we enter these realms, the information that is needed for us to heal becomes available to our conscious mind. By integrating this information fully, we create a new paradigm of awareness. What we call healing occurs as a direct consequence of the mind-body and emotional systems restructuring themselves to match or reflect this new level of inner awareness or consciousness, this new frequency."

One of her clients, Susana, was 67 years old and living in Argentina when Dr. Carbone started virtual sessions with her. Here's how Dr. Carbone describes their sessions together:

"Susana came to me as a referral from a colleague psychologist. She had been struggling with mental health issues for over 35 years and had been diagnosed with anxiety disorders and clinical depression. Her symptoms worsened with the passing of her husband, two years prior to our first meeting, and were greatly interfering with her day-to-day activities. These included regular panic attacks, social anxiety, a general lack of interest in life, and insomnia. She had tried many different therapeutical approaches, including psychotherapy and psychotropic medication. Additionally, Susana had chronic digestive symptoms and was recovering from lobular carcinoma.

"During our first consultation, I understood that, more than anything, Susana was in deep need of a 'safety net,' of validation, and of love. I decided to use my sound bowls to induce Susana into a relaxed mind-body state that would allow me to begin to work with her at a deep, subconscious level. During our two-hour session, Susana was able to access and integrate childhood and prenatal trauma that she had not even been aware of. There was a great emotional and physical release. She was exhausted and astonished by what we accomplished in just one session. We agreed to meet in one week.

"At our next session, the change was already visible: Susana's face was refreshed, brighter, younger; and there was a permanent smile. I welcomed her and she responded, 'What did you do to me? In all these years of therapy and medications, I never experienced anything close to what I was able to accomplish with you in just one week. I am a new person.'

"Susana continued to share that since our first session, she had been able to sleep through the night without medication, and two days later she decided to drive and run her errands alone for the first time in almost a year. A few days later, she went for a walk on the beach, and most of her digestive issues were simply gone.

"Susana became a regular, weekly client for about six months. We continued working with a combination of sound therapy and other therapeutic modalities. Now our sessions are more of a monthly check-in for support. Her symptoms continue to improve daily. Under medical supervision, she has significantly reduced her medication intake and,

> most importantly, she is able to enjoy her life, her family, and even go to social gatherings and take trips."
>
> Susana is a great example of what is possible from a whole-self therapeutic approach that redefines the traditional concepts of "normal," age limitations, and boundaries among the physical, mental, emotional, and spiritual realms.

MUSIC: A BALM TO EASE TRAUMA

In 2020, the Sacred Science of Sound presented a dynamic free, online summit. (All purchases of the summit after its completion went to support Minerva University scholarships.) It was a powerful time to gather in the spirit of music because of the political tensions, the Black Lives Matter movement, and the fears arising from the pandemic. The conversations featured eminent thought leaders and award-winning musicians such as Dr. Bruce Lipton, Dr. Levitin, India Arie, Victor Wooten, Dr. John Beaulieu, Eileen McKusick, Lee Harris, and Krishna Das, among others, and included the power of music to foster love and compassion.

I envisioned an interview with *New York Times* best-selling author Marianne Williamson on the vibrational frequency of prayer would be a good way to end the summit. As part of our conversation, Marianne began to share a story about a prayer-related incident that occurred when she visited a friend whose 12-year-old son had gone missing. The story was a humorous one with a happy ending about whether the mother chooses to pray or take a Valium.

She said, "To protect their privacy, I'm going to use fictitious names. So let's call my friend Linda, and we'll call her son Dylan."

She saw my face change.

"Do you know this story about the power of prayer?" she asked.

"No, Marianne, I don't," I replied, "but of all the boys' names you could have chosen, you've landed on the one that resonates the most. It belonged to my son." Then I told her about Dylan. She held me a moment in prayer. And we marveled at the mysterious and miraculous ways God works. This was no coincidence. I was being given a heavenly message he was with me, and a nudge to choose joy.

Joy was the resonant note for most of the speakers. Despite what was going on in the world, each one, in their own way, encouraged us to remember to find the small things that bring us Joy. I encourage you to find Joy as well. Joy running through our body affects our brain, our circulation, our autonomic nervous system, and our entire being, elevating us from a sense of anger and helplessness to hope and empowerment.

One of the most important experiences in my own journey of recalibration was a trip to Cambodia, where I was invited to teach for a program called "Cambodia Sings!" This is a non-profit organization that offers community singing to young Cambodians as a way to transform their trauma and inspire their creativity and confident self-expression. I went to train the teachers in my Whole-Person-Awareness singing technique, then taught it to the children and young people. We opened our voices, practiced breathing exercises, played creative games, and made memories. We sang the songs "Can You Feel the Love Tonight" from *The Lion King* and the spiritual "This Little Light of Mine." Everyone who wanted to do so took the microphone and sang solo. I received so many touching hugs of gratitude from the children both big and small.

When we open the human instrument with the exercises I used in Cambodia, we recognize how each person's voice is an integral part of the group and revitalizes the spirit. I also worked with a choir and coached the young adults attending music school in their native Cambodian songs and classical standards in German and French.

The national silencing and brutal killing of almost one-fourth of the Cambodian population during the time of the Khmer Rouge had left a deep legacy of fear in people regardless of their age, making it difficult for them to express themselves freely, whether speaking or singing.

During my time in Phnom Penh, the hardships this nation had gone through were with me. I wanted to reactivate the power of singing and share the healing sounds of the crystalline instruments. It was an honor to connect our hearts, and the sacred vibrations brought healing energy for all. Before I left Cambodia, I went to the Killing Fields where more than a million people had died. I held the pure frequency of intentional sound in my consciousness. There is a stupa there containing more than 5,000 human skulls of those who were murdered by the Khmer Rouge. I walked in silence, my heart in my throat. A wooden sign announced, "Killing Tree against which executioners beat children." There, on that tree, I saw many colored string bands displayed that were bright and hopeful. They are bracelets left there as expressions of love for those gone too soon, as a prayer, as a fervent wish for peace. I stood in silence. I felt my son so close and heard him speak. "Mom, it is horrific, but remember what we are here for. Music is joy and gives strength." I left his photo there at the tree as a gesture of guardianship, love, and care. This was a moment that will always remain with me.

As I walked out of the Killing Fields, I could not help but think of the concentration camps in Germany and Poland I had visited, and the piles of shoes—small, medium, large—engrained in my memory. Children. We all walk this earth deserving safety and love. Yet there is Pain. Loss. Despair. Violence. Families torn apart. The ancient balm of Sacred Sound Vibration and Music activates Hope and Healing. This, I know.

CHAPTER 6

AMPLIFICATION

The Science of Sound Healing

> The key to understanding how we can heal the body
> lies in our understanding of how different frequencies
> or "tones" influence our physical reality.
>
> — HANS JENNY, *CYMATICS: THE STRUCTURE AND DYNAMICS OF WAVES AND VIBRATIONS*

When working specifically with sound for healing, two different concepts are important, as my friend Jonathan Goldman explains in his book *Healing Sounds*. One is the frequency, the actual sound that is being created. The other is the intention of the person making and receiving the sound. He puts these concepts into the following formula, which I'm pleased to have his permission to share here:

Frequency + Intent = Healing

Jonathan and I speak often, and recently he reminded me that this principle is perfectly illustrated in the work of Masaru Emoto, who showed that crystallized water takes on either harmonious or warped shapes, depending upon the intention projected upon it. He continued:

> When a positive intention such as "love" was encoded in the water, these crystals looked like snowflakes. When

negative intentions such as "hate" were encoded, the water crystals looked ugly and muddy. When we think about how much of our bodies (approximately 70 percent) and our planet (about 71 percent) are composed of water, we realize the importance of using positive intention whenever we can.

He explained further that this ultimately means that the sounds we create, whether through our voices, different instruments, or different sound devices, ultimately impact the vibrations (the frequency) and the consciousness we project onto the sound (the intention).

We are at a crucial time in our development on Earth, with great suffering in evidence everywhere. Jonathan shared, "I believe it is possible to use sound to uplift the consciousness of all beings on the planet through intentional sound vibration. When we are in a state of coherence—where our hearts and our brains are resonating together in appreciation and gratitude—the electromagnetic field we produce is anywhere from fifty to five hundred (or more) times greater than normal. When we add the element of sound the field is amplified many times more. This is why the different prayers on our planet, regardless of the tradition, are chanted, whispered, spoken, or sung: sound focuses and amplifies the power of our prayers. When we gather to create sacred sound encoded with positive healing intention, and then project this energy outward to our planet, we are able to create a sacred, sonic field of sound that can literally shift and change our vibratory rate and raise the consciousness of all beings."

Jonathan has been an important sound-healing mentor and inspiration to me and to many. I integrate his formula, Frequency + Intent = Healing, in all of my work. Singing and toning, in combination with the sounds of the pure quartz crystal singing bowls—alchemized with gemstones, organic substances, and precious metals—permeates your senses and creates a soothing balm like nothing else.

HOW WE HEAL

Processing my emotions with crystalline sound brought insight and clarity. I recognized that Dylan was guiding me into a new purpose using the singing bowls in tandem with my music skills and felt the unexpected level of healing happening within me. It was time to form a company. I was inspired to create the Crystal Cadence Sound Healing Studio and Temple of Alchemy. Why that name? To illustrate structure. *Cadence* is a musical term for something that brings a phrase or a piece of music to a resolution or closure. I like to refer to a crystalline cadence as a pattern of sound frequencies that move us to a place of profound stillness, of deep rest, where the harmonics create a safe container for feeling, healing, integration, health, and wholeness. The structure of the cadence contains the vibrational wisdom of the ancient world presented to us through these astounding modern crystalline alchemy instruments.

Alchemy sound has been known to reduce sleep difficulties and help create emotional balance. This can relieve anxiety, stress, depression, addictions, and physical pain. The frequencies and harmonics of the quartz crystal singing bowls may clear negative thought patterns and behaviors and uplevel your neurology. The grounding qualities of Quartz may also enhance your connection to your higher self and activate embodied love. When we are in that vibration of love, everything is possible, and healing can happen. When we listen, breathe, receive, and integrate, healing can remain. The celestial sound of the singing bowls allows a true bridge from the physical to the spiritual connecting you to your soul, which has the Infinite as its reference point.

> ### Sound Rx
>
> Wendy Leppard, a graduate of the Sacred Science of Sound Crystal Alchemy training program whose experience with a worker at her home was shared in Chapter 1, offers a great example of the energetic connection the crystal singing bowls facilitate.
>
> One of her employees, Melody, was about four months pregnant when Wendy received her first three bowls, and as soon as Wendy began playing them, Melody's baby responded. Melody agreed to be the subject of a crystal bowl case study, and Wendy began providing weekly sound-healing sessions for Melody. The timing was determined by the baby, who Wendy reports would "call" both women to let them know when he was ready.
>
> "I would call Melody for a session, and I would find out that at the exact same time without fail, she would have had the 'quickening' too," Wendy reports. "The sessions generally lasted about three minutes, and then we both knew when her baby had had enough. I completed the sessions I needed for my case studies, but the baby decided that we would continue—which we did, right up until Melody went on maternity leave."
>
> Melody and Wendy had arranged that Wendy would play the bowls when Melody was giving birth. Melody's two previous births had been very difficult. Although Melody didn't let Wendy know when she went into labor, Wendy knew when it was happening, as the baby had "let her know," and she began playing the bowls.
>
> "As you can imagine, this birth was different!" Wendy notes. "Melody's beautiful Rowman was born after three hours of labor, which flowed smoothly and easily. The experience afforded to me by Rowman has been the most humbling and most inspiring experience of opening up to much deeper levels of communication that are available to us. It's time for us to listen deeply within, be curious and in wonderment, and continue to notice what we notice that is different in every moment."

THE MEDICINE OF TODAY

Today's medicine is acknowledging the fact that many ancient cultures used sound as a remedy for health. The fascinating studies of the late Swiss scientist Dr. Hans Jenny has showed the correlation of vibration into physical form in his book *Cymatics: The Structure and Dynamics of Waves and Vibrations*. The late Dr. Mitchell L. Gaynor, author of the books *Sounds of Healing* and *The Healing Power of Sound,* integrated crystal and Tibetan singing bowls in his practice with cancer patients as an effective, holistic approach to mind-body healing.

As I've mentioned previously, the human body is crystalline in structure: our bones are made of collagen and calcium phosphate, a mineral crystal. Our blood exhibits liquid-crystal properties, and our nerves generate electricity due to a liquid-crystalline effect in their myelin sheaths. Researchers at Imperial College London and the University of York have found that the "principal building blocks of crystals at this tiny scale inside our bones are curved, needle-shaped crystals that form larger twisted platelets to resemble propeller blades. They appear lacy and rodlike in their interweaving patterns."[1] And even the lens of your eye is actually focusing magnetism via the matrix of its paracrystalline arrangement of proteins that have a high degree of spatial order, similar to the matrix arrangement of atoms in clear quartz or lead-crystal glass.[2]

We are vibrating crystalline vessels, and therefore the crystal-clear frequencies of the singing bowls are not only beautiful to our ears, but they also penetrate deeply into our tissues. Acoustic-physics scientist John Stuart Reid, creator of the CymaScope (a device for visualizing sound) and a guest speaker on the Sacred Science of Sound educational platform, shared the following striking example with me.

John described that when a system of cells is challenged by a pathogen, the cells enter a sleep state known as the *G-zero phase*, resulting in the person feeling ill. Rather like the story of Sleeping Beauty requiring the kiss of the prince, sleeping cells can be reawakened by the "kiss" of frequencies provided by crystal bowls, creating sonic nourishment for our cells. And when the pulsating sounds of crystal bowls enter our bodies, they stimulate the lymph vessels with a gentle form of massage that helps eliminate bodily toxins.

John went on to note that the very low frequencies from the large, 16-inch Supergrade crystal bowls create an intense, shimmering wave sound beat, pulsing a steady, deep-toned frequency that stimulates the vagus nerve, reducing chronic inflammation and slowing the rate at which we age. The low-frequency sounds of the Supergrade bowls also innervate our sex organs. (Please scan the QR code or follow the link at the back of the book to experience these sounds.) John shared with me that the high frequencies from small crystal bowls stimulate production of nitric oxide in the paranasal sinus cavities, which reduces blood pressure and increases blood flow, bringing more oxygen to the tissues and powering our libido and many other important healing mechanisms.

In John's own words:

Bathing in the sheer beauty of crystal bowl sounds elevates our emotions. It transports us in our imaginations, boosting our dopamine levels—the happy hormone that supercharges our immune system—and reducing cortisol, the stress hormone. This all helps our bodies maintain a stable and optimum homeostasis. These are just some of the many ways that the exquisite sounds of crystal bowls support health and healing: they represent a powerful form of Music Medicine for the 21st century.

Crystals and Their Unique Powers

John has explored crystals in his laboratory and has arrived at many fascinating insights. For example, he shared with me that all human beings are powerful emitters of electromagnetism (light) and that people radiate infrared light (IR) with a power of approximately 100 watts (at rest) and up to 700 watts when running or engaged in extreme exercise. John also mentioned that when we hold a crystal or when we are close to one, the IR that we are emitting enters the crystal. The atoms in the crystal lattice are electrically charged, and when the IR light enters the crystal, it enters a lattice of vibrating electric charges and electric fields. The IR light's own electric and magnetic

> fields interact with the fields already present in the crystal, and the crystal then emits its own energy—which means that the person is exchanging energy with the crystal!
>
> John also shared with me that a similar process occurs with our heart energy. Humans emit a powerful, low-frequency electromagnetic (EM) field from their hearts. When the heart field enters a crystal, it interacts with the electric fields already present in the crystal, resulting in an energy exchange between the crystal and one's heart. This is why we can sense a crystal when we hold it in our hands or are close to it. John also talked about crystal excitation by sound and the fact that crystals have piezoelectric properties; therefore they can be easily excited by singing to them. Singing to a crystal causes small perturbations of the crystal lattice, resulting in the crystal radiating a varying electric field in sympathy with the sound field, which has healing potential if the crystal is placed on a client during the vocalizations.
>
> Regarding the sounds of the crystal singing bowls, John confirmed that they are powerful emitters of both sound and infrared light, and that the IR light they create is modulated in amplitude and frequency by the bowl's sound. Here's an astonishing fact: since the cells of the body communicate with each other primarily in the IR spectrum, the sounds of the bowls are, in a sense, speaking the language of cells!

I am honored to share the voice of another friend of the Sacred Science of Sound who was living near Silicon Valley at the time the crystal bowls were being used in the computer industry and birthed as instruments.

Dr. John Beaulieu, one of the major innovators in the world of sound-healing therapy, is a world-renowned speaker, composer, pianist, and naturopathic doctor. John pioneered a technique called BioSonic Repatterning and co-founded BioSonic Enterprises Ltd., a company dedicated to "tuning the world."

Dr. Beaulieu spoke at the Sacred Science of Sound online summit in 2020, and we laughed about his experiences with both the quartz singing bowls and tuning forks. In the 1980s he had a collection of large quartz singing bowls that he used in his music and played live in his concerts. However, those original classic frosted singing bowls were big, bulky, and heavy, while the tuning forks were practical, small, and lightweight. He smiled as he said he could safely tuck them in his back pocket. The forks were his on-the-go sound-healing instrument of choice—much easier to transport than the singing bowls! (Oh, how I can attest to this!)

He shared how after 9/11, he was called to supervise a group of his trained therapists who were giving sound-healing sessions to firefighters working at Ground Zero. In an improvised clinic right in a Lower Manhattan fire station, the therapists would tune firefighters with biosonic tuning forks as they walked in from the disaster zone, still wearing their gear. They used the notes of C and G—the musical interval of the perfect fifth, which is associated with relaxation—in combination with the Otto tuning fork, another C note vibrating at twice the frequency, to help these incredibly stressed firefighters return to a state of inner harmony and neurocoherence. Dr. Beaulieu humbly shared that this initiative helped bring sound healing into the mainstream.

> ### Sound Rx
>
> Dr. Maja Jurisic, a physician and graduate of the Sacred Science of Sound Crystal Alchemy training program, sees tremendous therapeutic potential in the crystal singing bowls:
>
> "As a physician, I became interested in sound as a way of helping people to heal when I realized that traditional Western medicine pays lip service to the need for holistic treatment but still treats patients' bodies and minds as separate entities. Bodies are viewed mechanistically. Treatment plans are devised to "fix" what is perceived as broken rather than treating mind, body, and spirit as inextricably linked so as to arrive at coherence and wholeness. . .
>
> "After taking several of Jeralyn's classes, I started playing my bowls for family, friends, neighbors, and colleagues who were dealing with challenges in their lives (myself included). They invariably found it relaxing and beneficial. One of my friends shared that the sound bath I did for her helped her shed the anxiety she had been carrying around since she had been diagnosed with breast cancer.

Amplification

"Most recently, I did a virtual sound bath over Zoom for a dear colleague diagnosed with kidney cancer and scheduled for surgery a few days later. I introduced her to the bowls I would be using, explained I would be toning at times with the bowls, and primarily playing musical intervals of fourths (which calm the amygdala, our emotional fear center), fifths (which have a balancing effect on our nervous system, as well as increasing nitric oxide in our bodies, which supports the immune and cardiovascular systems), and octaves (which are grounding and calming). And I invited her to set an intention for the session.

"As I started playing, I noticed that her features relaxed. I was primarily playing a third-octave F bowl with Selenite and Saint Germain, a Quartz middle-C bowl, a fourth-octave F bowl with salts from the Great Salt Lake, an A bowl with Ruby, and a White Quartz high C. My bowls are all tuned to 432 Hz, which helps create a feeling of peace and, as many say, that tuning is aligned to nature. This combination also creates the sacred intervals of the octave, perfect fourth, and perfect fifth Jeralyn so clearly teaches us.

"A smile appeared on my friend's face as I played and toned. But after 10 minutes or so, as the C bowls' sounds faded, I then started playing a D bowl with added Sandstone from Sedona and Platinum. Her face and posture then further transformed, tears streamed down her face (with the smile still in place), and she absolutely glowed. In the bowl-chakra system most commonly used in the United States, the D corresponds to the sacral chakra, which includes the kidneys. She didn't know which note I was playing, but her body did, and it responded.

"Afterward, my friend shared that as soon as she heard that new note, she felt something shift and could sense tingling throughout her entire body. As we said our goodbyes on Zoom, she looked more radiant than I had ever seen her and told me she had a very positive feeling as to how things would go for her.

"Even though I was treating her as a friend, not as a physician, I was moved at what a difference a 25-minute sound journey could make to someone's state of mind and what a profound effect sound and vibration could have on someone's body within minutes. I look forward to the day when this type of sound therapy will be offered by the medical establishment and accepted by insurance companies as appropriate treatment for any patients who need to retune, refresh, or restore themselves on their journeys to well-being."

THE CYMASCOPE: A BRIDGE TO AN UNSEEN WORLD

To bring another perspective to the conversation around sound, vibration, and healing, I asked John Stuart Reid to explain the CymaScope instrument and why it is important for the continued development of the science of sound healing and music medicine.

John began his reply by pointing out that with the exception of music and singing, many man-made sounds, such as engine noise, can be jarring, while the sounds of Nature tend to flow over and around us like soothing waters—lifting our spirit, inspiring us, exciting us. Yet if we could see sound, our world would be even more beautiful than we could imagine. It would be a world filled with shimmering, holographic bubbles, each displaying a kaleidoscopic pattern on its surface. To see sound is to open a new window onto our world, and when we can see something, we can understand it at a far deeper level than with our other senses.

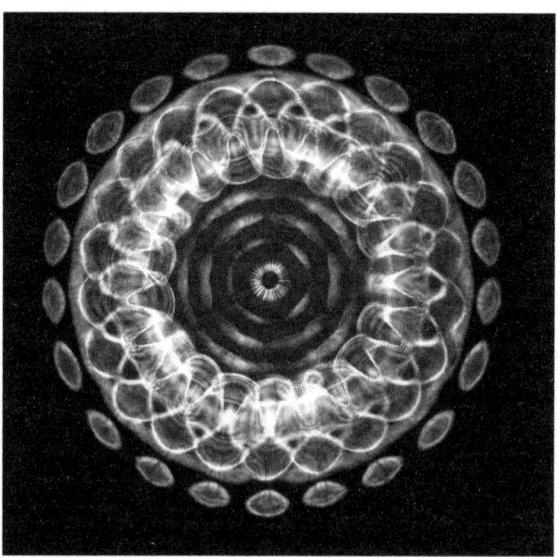

The CymaScope makes sound visible by imprinting sound's invisible vibrations onto the surface of ultrapure water. Here, we see an F note played on my 432 Hz–tuned, six-inch Charcoal Palladium bowl with an unusual visual expression of 22 antinodes.

Amplification

The CymaScope instrument John invented allows us to see a previously invisible realm—the world of sound and music—helping us to gain a deeper and fuller understanding of frequency, vibration, life, and the Universe. He explains:

> The CymaScope is a calibrated scientific instrument that makes sound visible by imprinting sound's invisible vibrations onto the surface of ultrapure water, creating a geometric analog or model of the sound, similar to making a fingerprint visible on glass. In the case of a fingerprint, the surface is dusted with fine powder—the revealing medium—while with a CymaScope instrument, the water is "dusted" with light. The resulting "CymaGlyphs"—another word for "sound prints"—can be analyzed to discover aspects of the originating sound that are not revealed by conventional scientific instrumentation.
>
> Looking at the CymaScope image above of the octave-5 F note I played on a 432 Hz–tuned, six-inch Charcoal Palladium bowl (see page 78, and also the cover of this book) we can understand that as the bowls are being played, these perfect visual images are vibrating in the fluids that compose our body, forming sacred-geometrical patterns that help bring harmony to our body chemistry. In John's own words, "This includes increasing the production of leucocytes that strengthen our immunity; reducing the production of cortisol, the stress hormone; and activating dopamine, our happy hormone—bringing higher levels of health and well-being."

John notes that sound lies at the heart of every aspect of Nature, and the CymaScope opens a bridge to an unseen world and to the Universe, a bridge that was not possible with previous technologies. He also mentions "the ability to see sound opens the way to significant advancements across many scientific disciplines, because sound underpins almost every aspect of creation. And indeed, the CymaScope can make visible the sounds of all aspects of our physiology, such as the sound of heartbeats and the sounds generated by every cell. The discovery that cells create sound was poetically termed 'the song of the cell' by Professor James Gimzewski, who coined the word *sonocytology* to refer to the study of sounds created by cells."[3]

CYMATIC PRINCIPLES

According to John, the underlying principle of cymatics is simple:

> When sound encounters a membrane, such as latex, the surface membrane of your cells, or even the surface of a thin metal plate, the molecular vibrations of that sound imprint a geometric pattern upon the surface that is representative of that sound. Such patterns are normally invisible, but they can be rendered visible in several ways. For example, in the case of a thin metal plate that is excited by sound, the invisible pattern can be made visible by sprinkling fine powder or sand on top of it. The powder will gather in regions termed *nodes* where the plate is not vibrating, while the *antinodes*, which are the regions of maximal vibration, push the powder away.

This principle is also at work in the fluids inside your body when you are immersed in sound.

The image of the multicelled CymaGlyph John has so kindly allowed us to reproduce was created with fine quartz sand on a circular brass plate with a pure, single frequency. The striking similarity to the structure of a diatom, a sea creature that first appeared in the oceans of the Jurassic period around 185 million years ago, points to the intriguing possibility that the morphology of early life-forms was influenced by the presence of ocean sound, or even more fascinating, the possibility that sound triggered the earliest stirrings of life. John poses the interesting question: "Sound is an aspect of life, but is life an aspect of sound?" John went on to explain potential sources for this life-forming sound. "The sound that created life could be the jostling of bubbles from hydrothermal vents on the ocean floor, creating a kind of white noise, or the complex sounds of wave action, which are naturally filtered as they penetrate the ocean depths, creating different frequencies at different depths. In accepting this concept of sound creating and shaping life, it becomes obvious why sound can heal life: its vibrational frequencies catalyze the body's many innate healing mechanisms that lead to wholeness and health."

A complex, multicelled CymaGlyph created by powder on a sonically excited metal plate.

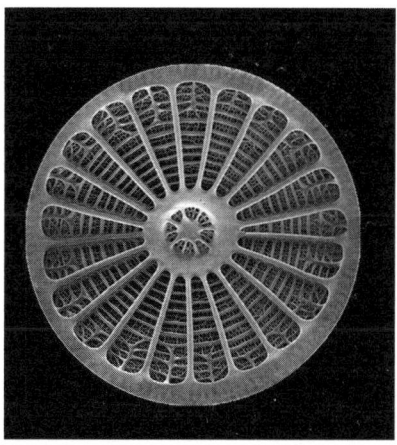

The diatom genus *Arachnoidiscus* first appeared in the Jurassic period. Note the striking similarity to the image of a complex, multicelled CymaGlyph created by powder on a sonically excited metal plate.

It's compelling to contemplate that sound created life and that those ancient life-forms resemble sound made visible. (And it's no coincidence that in addition to sharing our love of music, Dylan and I collected ammonites and fossils!) As we reflect on the history and universality of sound, we return to the concept of that pivotal moment: that in the beginning was Stillness. Sound Vibration. Birthing. Infinite. Majestic. Profound. The Tone of Creation. The Music of the Cosmos, the Universal Key. Divinity. The Frequency of Love. God. And as Dylan always reminds me, "Mom, there is nothing bigger than that."

CHAPTER 7

ELEVATION

The Quantum Is Real

Grapes must be crushed to make wine. Diamonds form under pressure. Olives are pressed to release oil. Seeds grow in darkness. Whenever you feel crushed, under pressure, pressed, or in darkness, you're in a powerful place for transformation and transmutation.

— **LALAH DELIA,** *VIBRATE HIGHER DAILY*

One day in mid-May 2015, six weeks after Dylan's passing, I received a strong message from my son: "Mom, it's time. We are going to Machu Picchu."

It had been on our wish list, and we had not made it there together.

"Okay," I said, and asked him to show me who we were going with. I opened my computer and searched for "spiritual journeys to Machu Picchu." The name "Dr. Sue Morter" came up as the first entry. I did not know her, but I saw she was taking a group to Peru in three weeks' time, and I knew I needed to go. Although I was imagining being guided by a native Peruvian shaman, I couldn't find any opportunities that aligned with that desire.

Dylan finally said loud and clear in my ear, "Mom, stop googling! Dr. Sue is the shaman, and we are going with her!"

When I shared this with her, she laughed and told me she was nicknamed "the shaman in the business suit."

With anticipation and excitement, I booked my ticket to Peru, and when I finally landed in Cusco and arrived at the hotel, I felt nervous and alone. The high altitude made the air thin, and I had to use an oxygen machine that was available to hotel guests to calm my breathing and fall asleep. The next morning, I entered the orientation room, and as soon as Dr. Sue walked in, tears rolled down my cheeks. It was a clear recognition that I was exactly where I was supposed to be and that on some level, I already knew her. I was showered with kindness and embraced by everyone in the group.

What happened in Machu Picchu changed the way I looked at everything. On one of the first days, we visited Sacsayhuamán, an Incan fortress-temple, and I remember standing at that power spot while Dr. Sue introduced the concepts of the Life Purpose chakra, the Soul Purpose chakra, and the Legacy chakra, three energy centers outside the physical body. As I stood at the monument with my back against the ancient stones, a strong physical sensation of stability and connection to the earth began to anchor me to my bigger purpose. I was embodying a sensation of rootedness and a flow of energy throughout my body that I had never felt before. I had a profound awareness of my soul connection with Dylan and the reasons he had left so early. This sense of Dylan's presence alongside the physical experience revealed an unprecedented clarity within me, a heightened state of consciousness. I was speechless, in awe of what was being revealed. And everywhere we went on that trip, Dylan was with me, laughing with joy at the llamas and being part of my adventures.

When we arrived at our hotel in the Sacred Valley, the place that was to be home for the next few days, each room had a name on it, and I was astonished to see that the nameplate on my door read "Tintin." I gasped. How had they known to give me *this* room with the name of Dylan's beloved comic book and its leading character? At that moment I heard Dylan laughing. "Well, Mom, are you having fun yet? We came all the way to Peru to meet the Spanish-speaking Tintin!" I had come to Peru vulnerable and lost, and of all the rooms available, I had been given this one! What's more, I learned that *tin tin* meant "passion fruit" in the native Quechua language, and it was one of our favorite fruits. These synchronicities gave me comfort and confirmation that I was on the right track.

The name of my hotel room at the Willka T'ika Hotel, the same as Dylan's beloved comic and leading character by Hergé and also native Quechuan for a favorite snack, was a comfort and confirmation that I was on the right path in Peru. (Photo by Jeralyn Glass)

And there the most extraordinary thing of all happened. On day three, our group was to get up at 5 A.M. for the strenuous hike to the top of the iconic mountain, but I was in a weakened state, as I had spent the previous night weeping. I declined to go. I chose instead to wander across the valley and spend time by myself walking at the Sun Gate, an ancient archaeological complex built of terraces, stairs, and buildings. It is a spectacular area where the sun illuminates the main gateway during the winter solstice. Throughout the day I would gaze up at the famous Huayna Picchu, where I imagined the rest of the group to be.

When we all reconvened on the train to head back to our hotel, Dr. Sue, with tears in her eyes, said, "There is something I must show you."

She told me that when the group arrived at the peak, she created a special ritual for Dylan, which they had photographed and filmed for me. While standing on the mountaintop, Dr. Sue read the quote from Dylan's college application that was on the back of the photo of him I had given her when we met:

> My experiences on the mountain drive me forward, making me want to succeed, to climb up the mountain of difficulty, conquer its peak, and enjoy the ride down. Although the ride down is exhilarating, the hike up is just as important. As Heraclitus wrote, "The way up is the way down." I value every moment of my time on the mountain, and every moment of my life.
>
> — *Dylan Sage*

Then she took the photo and "showed" Dylan the spectacular 360-degree view from the top. ("As if he couldn't see it already," she said with a wink.)

When Dr. Sue Morter "showed" Dylan the view from the top of Huayna Picchu, a radiant beam of light landed directly on his throat in a miracle of energy and light. (Photo by Greg de Haan)

A radiant beam of light came in from above, wobbly at first, then becoming stable, and it landed directly on Dylan's throat as a golden ray. The miracle of energy, light, and eternity is fully represented in this image, and everyone who witnessed this moment was brought to sacred stillness.

Peru embraced me in my sadness and opened the path of healing. It taught me that although we change forms, our essence never dies. (Please follow this link to hear an aria accompanied by strings and crystal singing bowls and to see the miracle of the light on Machu Picchu: crystalcadence.com/sacred-vibrations-content.)

I was astonished by all I experienced on that trip and grateful that I no longer felt isolated and alone in my grief. After returning home I fully immersed myself in Dr. Sue's teachings. During this time, I finally began noticing moments in which the pain felt less heavy. Her work gave me a foundation in energy medicine and a grounded understanding of the science behind the healing power of sound. All this provided a reason to live, fully immersed in this ever-expanding world of energy and sound medicine. The study, the practice, and the integration of the principles of bio-energetics became my go-to tool kit to process, reclaim, and recycle stuck energies. This helped me commit to my personal healing process and honor the eternal soul connection with my son, bringing Heaven to Earth. Miracles filled my life and became the guiding light on the path to recalibrating my pain.

Dr. Sue invited me to share my crystalline music with her community. I began to play at her events and accompany her free monthly Transmissions with exquisite soundscapes. The power of the multilayered celestial harmonics in combination with Dr. Sue's presence gave me and thousands of others a sacred container to breathe, feel, and heal.

Dylan gave me a heavenly nudge to connect with Dr. Sue Morter. She was present at exactly the time that love and unconditionality were most needed; holding me, guiding me, and teaching me as I traversed the greatest loss I could have imagined. She showed me a way to understand my soul contract with Dylan, or as she calls it, "the bus-stop conversation"—the idea that before coming to this life, we have a soul agreement about the different roles we will all play in each other's lives. And this consensus happens humorously, as we're sitting together at "the bus stop." Although I was hesitant to accept this concept, I began to understand its relevance, and this was made evident by looking at the synchronicities in our life together from a higher viewpoint. Still, there were moments I would protest, and look at Dr. Sue with great skepticism, questioning how Dylan's early death could possibly bring anything good.

Then one day I remembered a very poignant message that I had received barely two weeks after Dylan's passing. I was on a group call with a psychic who spoke directly to me. She said pointedly, "You know that Dylan was never going to live a long life, don't you?" Shivers went through me as she spoke clearly, and evidently confirming what some part of me knew. She continued, "Whether it was now, or at 22 years old in a car accident, he was never destined to live into adulthood." I was overcome with emotion as I reflected on the truth of that statement, recognizing how much we had packed into our 19 years together. In the deepest part of my being, I knew this was accurate. The concepts of the bus-stop and bird's-eye view of life were becoming clear.

WE'RE GOING TO DO SOUND HEALING

One day in August, about five months after his death, and a few months after my trip to Peru, Dylan spoke to me as I was sitting in the kitchen. We always loved cooking together and hanging out in the kitchen, and that hasn't changed. Some of his most powerful communications have come through in the kitchen.

"Mom," he said, "Call the bowl dudes. We're going to do sound healing. . . "

"Huh? Son, I'm grieving you," I said. "Please, leave me be!"

I then ignored him. I laugh about that now, as he truly would not be ignored and did not leave me be. Dylan was large in life and larger in Spirit. My conversations with him are engaging and often quite humorous. Every day, when he repeated, "Mom, call the bowl dudes. We're going to do sound healing!" I would just tell him to forget about it. I wasn't interested. But he wouldn't stop pestering me about it, and finally I made the call.

I shared that I was a professional singer and musician and that my son had recently passed away. I had a set of seven bowls, and we had started playing when he was seven years old. Now, from the heavenly realms, he was instructing me to start a grounded path in sound healing with the crystal bowls and my voice. I then chose 11 bowls with the intention of using them to ease my grief.

I chose strictly by alchemy, not by note. I was more attracted to what the alchemies evoked. What might help connect me to Dylan? Ruby for transformation. Selenite for grounded white light and connection between Heaven and Earth. Charcoal for clearing and detoxifying. Rose Quartz for the heart. Celestite for contacting the celestial realms. Amethyst for bliss and spiritual awakening.

When my bowls arrived, I sat with them all laid out in front of me and began to play. I went slowly, gently tapping to hear the note and then swirling the wand to land in the overtones. Some resonated with me more than others. I was extremely drawn to the Selenite bowl, and I began there. As I played it, I saw a misty white substance fill the room. It felt like angels. Then I became aware of a presence. Dylan seemed to be guiding me with that bowl. I felt him showing me what to do. I began to tone with my voice as I played the bowl, and I expressed sounds I had never made in my life: heart-wrenching groaning, keening, screaming, protesting, and anguishing sounds. Resolutely embraced by Dylan's Spirit and held by the grounded white light of Selenite, I released and let go. Sometimes the bowl screeched and the sound was excruciating, like nails on a chalkboard. I recognized later that the sound was simply mirroring the pain and grief inside me, as three days after that, to my great surprise, the instrument sounded normal again. What I have come to understand is that the Selenite bowl activated the area behind my throat, the zeal point chakra, sometimes referred to as the "mouth of God." The combination of the note and the alchemy of that bowl helped create an easy pathway to connect with my son and to receive communication and support from a higher dimension. Feeling held and protected, I dared to express the messiness and depth of feeling I could not reach in talk therapy. I played for almost an hour, lost in sound, claiming, then releasing, what I had judged to be too big, too impossible to feel, all the while knowing that in the enormity of those emotions, I was safe, and I was healing.

We want to avoid thinking about death and feeling grief at all costs. Add in shock, trauma, and remorse. It's not a great combination. How did I let these go and find peace?

While my voice entrained with the Selenite bowl, I gave myself permission to breathe and feel. I will always remember getting up to go wash my face after that hour of playing, and as I saw my reflection in the mirror, I also saw there was light in my eyes, a light I had not seen since Dylan's passing. What was happening within me? I saw a spark of joy reflected back to me. How was that even possible? And why was this experience so different from talk therapy? The sound vibration gave my mind something to focus on which was closer to my soul, and my mind was no longer clamping down and compressing me with grief. Playing the Selenite bowl opened internal space. I could actually feel the energy unabashedly moving through my body with full permission to be felt. Through vibration, my mind quieted and became a healthy *participant* in facilitating and enhancing healing.

As a singer, I knew at the core of my being that sound vibration and music take us beyond words, beyond human touch, into the place of the ineffable. I experienced this regularly in my life, whether I was training my voice, singing performances, or as a concertgoer or music listener. Yet what happened that first night with the new alchemy bowls was different; this was real life with raw emotion being expressed in a situation that was unchangeable, and when the door to experience the healing power of crystalline sound opened, Dylan was with me, guiding me and supporting my transformation.

Big Angel, I look to the heavens and there you are. I land inside myself, and there you are, anchored in my heart, a part of me, inseparable. Everywhere and Nowhere. An Invisible Presence made up of Sacred Vibrations.

CHAPTER 8

REVELATION

Signs of Synchronicity in the Himalayas and Beyond

There are only two days in the year that nothing can be done. One is called Yesterday, and one is called Tomorrow. Today is the right day to love, believe, do, and mostly to live.

— HIS HOLINESS, THE DALAI LAMA, *HEART TO HEART*

As part of my ongoing journey to wholeness, I made plans to attend a retreat in India in the Himalayas hosted by mantra singer Kevin James. Dylan and I had been to India when he was 17. We had also visited Nepal and had had the good fortune to see Mount Everest by airplane. This famous mountain had been hiding behind clouds for days, and the morning we went on our airborne excursion, the skies opened and that majestic peak revealed itself. We were overjoyed! We both loved the elevated views and the peace the mountains brought. Seeing Mount Everest had been a shared dream, and our plan had been to hike to its base camp when Dylan finished high school. We never made it there together: a friend's son, however, kindly offered to carry some of Dylan's ashes to base camp, and at 17,598 feet he gently and respectfully spread them on that sacred mountain known as the "Holy Mother."

This time, without Dylan sitting by my side, the flight to Delhi seemed unbearably long. From Delhi, I took a smaller plane north to the little mountain town of Leh in Ladakh. As we flew above the Himalayas, I saw something I had never seen before. Far away and very faint, to the left side of the aircraft, there appeared a round rainbow with the dark shadow of the airplane at its center. It was a fascinating sight. I had never seen a round rainbow, or the shadow of an airplane, during a flight. The round rainbow with the silvery shape of the plane in its center came more into focus as we flew over the famous mountain range. A group of photographers happened to be on the plane en route to a photo safari, and they pulled out their cameras and began photographing the rainbow out of the small windows of our aircraft. I was seated on the aisle, and the man next to me had a top-brand camera with a powerful lens. He began showing me in close-up clarity the rounded-rainbow pictures he was shooting. They were stunning—truly gorgeous—and he sent me his photos after we arrived in the village of Leh, high up in the Himalayan mountains, where the retreat was to take place. I also took some photos on my phone.

This trip to India came on the heels of the last Healing Sounds Intensive, a nine-day sound healing training in Loveland, Colorado, facilitated by sound-healing pioneer Jonathan Goldman and his wife, Andi Goldman. There, under their extraordinary leadership, I landed in a community of like-minded seekers and explorers of sound. Unbeknownst to me when I registered, that experience, for many reasons, was to become an incredible support for me to work with the deeper layers of grief that continued to emerge and that were longing to be felt and transmuted. Jonathan and Andi and their guest teachers created a safe, supportive, and pristine environment to be fully immersed in the experience of healing sound.

We explored sacred geometry with sound, and we were invited to enter—and tone inside of—a series of life-sized geometric figures shaped like Platonic solids. Sound invited us to connect with forgotten inner dimensions and embrace an ancient remembering of how human beings once used sound as a regular health tonic. We sang sacred songs daily, and on one day, the "Day of

Toning," we toned vowel sounds, mantras, and harmonics from 9 A.M. until 9 P.M., a full 12 hours of uninterrupted transformational vibration!

It was an intense time of investigating and receiving sound as medicine. A few of us walked to a nearby chapel and shared our voices with a choir of crystal bowl sounds. There was an enormous amount of Light available through sound that I had never experienced! On my last day of the Healing Sounds Intensive, I had a powerful and integrative private session with Nasiri Suzan, one of the co-facilitators. I remember telling her I was feeling vulnerable to be leaving immediately after Colorado to go to India, although the trip had been confirmed for almost a year. She encouraged me not to be concerned, that all was aligned for my highest good. She confirmed that I was supposed to join the Heartsongs in the Himalayas healing retreat with Kevin James, as planned, and assured me that later, I would be grateful for my courage and understand the reason why. I was open, and my heart was uplifted and tuned from the rich days of the Healing Sounds Intensive.

I flew home, unpacked from Colorado, repacked for India, choosing a very special set of four harmonic singing bowls to join me, and flew east to the Himalayas. When I arrived in that little mountain town, it was dusty and dry, but I felt oddly at home. I found there was something familiar and peaceful about Leh. I walked down the narrow streets and saw the first welcoming sign: there, hanging right in the front display outside a souvenir store, was a blue T-shirt with the images of Tintin and his little dog, Snowy, on it, along with the words "Tintin in Ladakh" embroidered upon it. Dylan had owned copies of the book it referenced—*Tintin in Tibet*—in both English and German. I could not believe the synchronicity of the fact that the Tintin T-shirt was one of the first things I saw when I arrived in the province of Ladakh!

And then I heard Dylan's voice through vibration, just as I had heard it in the Sacred Valley at Machu Picchu!

"Welcome to Ladakh, Mom! Are you having fun yet?"

An early sign of synchronicity in the Himalayas: A T-shirt featuring Dylan's favorite literary characters, Tintin and his little dog, Snowy. (Photo by Jeralyn Glass)

Dylan's sense of humor has always made me laugh, and it has often caught me off guard. But there was more: as I arrived at the square and waited for a taxi to pick me up for the hour-long drive to the retreat center, I looked again at the photos from the airplane. I gasped. Dylan's face was apparent in the center of the round rainbow! The tears rolled down my cheeks as I gazed unbelievingly at the image.

A beautiful photo portrait of Dylan wearing his favorite sunglasses and tilting his head slightly downward was made for his celebration of life. In the photograph from the airplane, that face—angled down with the sunglasses—was clearly represented at the center of the image of the round rainbow. I was in awe and filled with gratitude. That moment of connecting the dots, with Dylan's face in the middle of the round rainbow high above the Himalayas, activated within me a new sense of hope and infinite possibilities. I shared the photos with Kevin's group: my heart soared in wonder and joy! My time in India was filled with miracles, activations, integrations, and healing through sacred sound and mantra.

When I returned from my three-week trip to India, I sent the photos to Nasiri Suzan, the sound practitioner from the Goldman's Healing Sounds Intensive, and then called her. She laughed knowingly as she heard my voice and thanked me for the photos. Then she went on to tell me that Dylan had been very present during the private session I had had with her, talking loudly in her ear. He told her to tell me it was important that I go to India. He then said to her, "I will give my mom an unmistakable sign," and he instructed her not to tell me. I was speechless. That boy! A round rainbow—indeed, with his face in the middle of it? Truly an unmistakable sign.

The beautiful portrait of Dylan wearing his favorite sunglasses. (Photo by Dave Gregerson)

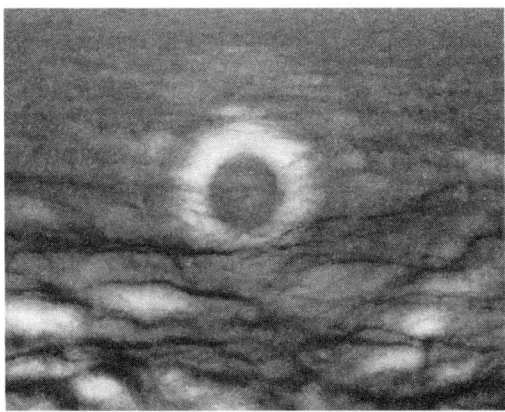

Recognizing Dylan's face from the portrait by Dave Gregerson in the middle of the round rainbow high above the Himalayas activated within me a new sense of hope and possibility. (Photo by Jeralyn Glass)

The journey to India revealed other signs. We hiked to an ancient Buddhist cave where we sang mantra, including one of my favorites of Kevin's called "Dance with the Whales." We prayed, and I left Dylan's photo high in the mountains in that sacred place. Three years later Kevin returned there, only to find Dylan's photo exactly as we had left it—smiling his inimitable smile. Life in human form has its way of challenging us as we contract and expand, uplevel and ground and we anchor into our authenticity and our Light. In the end, it's all there is. Energy, Sound, Vibration, Authenticity, Light, and the Frequency of Love. Love is the vibration that weaves its threads into our lives in a tapestry of miracles if we can only be awake enough to see them and be open enough to receive them.

The events of my visit to Ladakh served to support the idea of our undeniable soul contract and the bigger purpose to Dylan's early death. Every day Kevin led us in a morning practice that included yoga, meditation, and chanting. Every evening we sang mantra in sacred circle, sometimes in the local Monastery. During the day, we hiked, explored the land, interacted with the local people, and sang some more. One day we received information that His Holiness, the Dalai Lama, would be in residence in Leh, teaching publicly and giving Darshan, an experience of grace and connection that happens when seeing a holy being.

Our group was excited, as this opportunity was rare and unexpected. There were to be visits to the Thiksey Monastery on a Thursday and a Friday, and we were invited to choose which one we wished to attend. I chose Thursday. On Wednesday, I received a message from Dylan loud and clear. "Mom, go in the Friday group." And so, I asked to change. The Thursday group neither saw nor heard His Holiness and came home disappointed. On Friday we not only saw him and heard the translation of his teachings, but I stood three feet away from where he passed by with his entourage. Our eyes caught. He waved as if waving right to me. I could hear Dylan's laugh.

"Well, Mom? Got you a front-row spot to see His Holiness, and he waved right at us!" That boy! There were no words. I was speechless, filled with reverence and awe.

Six years later, one of my students from Malaysia had an audience with the Dalai Lama. She wrote to me about the experience and sent me a photo of His Holiness playing the large Platinum bowl she had purchased and gifted to him from the Crystal Cadence Sound Healing Studio and Temple of Alchemy in Los Angeles—a bowl that had sat on the shelf right beside Dylan's Saint Germain bowl. That boy sure has connections!

Dylan arranged for my awe-inspiring close encounter with the Dalai Lama. He waved to us! (Photo by Jeralyn Glass)

A HIGHER VIBRATIONAL FREQUENCY

In the pain of grief, there were gifts. Out of the anguish, blessings arose. After I came home from India, I was able to sustain the hope I received there for a while; however, grief ebbs and flows, comes in waves, and then recedes like the tide. Grief is a very personal journey, and there is no one right way to grieve. Each one of us must take our time, walking our unique path and navigating it to the best of our abilities.

One day I was going through Dylan's room, reliving memories, and was so overcome with sadness, disbelief, and anger that he was no longer alive, I lost it. I raged on myself and started pounding my chest as hard as I could, trying to beat the pain out of me. I made myself black and blue. I thought about ending my life to stop the intense and seemingly unbearable pain in my heart. In that moment my rational mind became stronger than the anguish. I came to my senses. I picked up the phone and called my mom, weeping, trying to express the intensity and depth of my feelings. She knew in an instant exactly where I was emotionally. She is a small but mighty lady, barely five feet three inches and 130 pounds at most.

"Jeralyn," she said to me, "for God's sake, I understand, but if you are really thinking of taking your life, I'll kill you first!"

Her words were exactly the sobering medicine I needed. I felt her seriousness, sharp and focused like a laser, a passionate cry for me to wake up and to become aware of what matters most. She felt powerless to help me and was struggling herself with grief. She had been so close to Dylan, her beloved grandson. I never fell to that depth of despair again. By allowing myself to express all the feelings I thought were unacceptable, frightening, and what I considered shameful, they released. Through everything I learned and practiced at the Healing Sounds Intensive and the HeartSongs retreat in India with Kevin James, where we chanted high up in the Himalayan mountains, another layer of cleansing and clearing grief had happened. I was being tuned. And Sacred Vibration was indeed the Medicine.

"I Got You, Mom."

We've had a family tradition, since my dad passed away in 2002, of spending the week following Christmas together on the Big Island of Hawaii, which my father loved. Dylan had been a part of that tradition since he was two years old. We made so many wonderful memories there, swimming with the dolphins and snorkeling with turtles. In the year after Dylan passed, I was swimming at our favorite beach on the Big Island. I had just seen two magnificent turtles swimming underneath me, and as I floated, dreaming in the warm ocean, I drifted away and my prescription sunglasses slipped off my head and into the water. I was to leave directly from Hawaii two days later for my first visit to Denpasar, Indonesia, and then go on to Thailand, where I would meet Kevin James and his daughter.

Feeling panic, I asked Dylan to help me find them. They were my only prescription sunglasses, and I would need them on the trip. "Please, son," I begged, "please help me find them."

> It seemed hopeless, as the water was murky and the shore waves churned. I got out of the deep and stood knee-high in the ocean, looking, hoping to see them emerge.
>
> "I got you, Mom," I heard him say.
>
> The ocean retreated in its ebb and flow, and there they were. I grabbed them out of the water and sand and vowed to keep them safe. I thanked my big Angel for protecting me in my fragility. Here was another miracle, more evidence of celestial support, of a connection beyond what my logical mind could understand or explain. I was leaning in.

FAR AND WIDE AND INTO THE DEEP

These travels out of my comfort zone were crucial for my healing, and I was completely supported by my big Angel guide, who was making sure this path of healing could unfold.

Kevin had suggested I experience a Chinese-medicine-based, award-winning retreat center at a spectacular property in Thailand called Kamalaya where he was engaged to share his healing music and lead chant circles with the guests. I checked online to see what other healing retreats in that country's general direction might be interesting. I had never been to this part of the world. I had been invited to sing in Japan, which I loved, but this part of the East was unfamiliar.

I landed at a yoga center in Bali, where I stayed a few days to ground and prepare for the adventure to come. Dylan was agreeing, laughing and guiding me.

"Mom, we saw the little dragons in the Galapagos"—he was referring to the iguanas—"now I want you to face your fears with the big dragons in Komodo National Park! I'll be right by your side."

He was serious! And so "we" planned a healing journey to what seemed like the far ends of the earth. And it was. After four days of yoga and detox in Bali, I flew—on a rather small aircraft—with four bowls from Denpasar to the island of Flores, where I would take a boat from the port of Labuan Bajo for my adventure to the Komodo Islands. We left the harbor promptly at 5 A.M. and went first to the smaller island called Rinca. There we had our first encounters with the dragons, huge lizards that can grow as big as 12 or 13 feet. They are carnivores that look like dinosaurs with large, forked tongues. And they are scary, sinister looking, and dangerous. I went with a guide who kept a stick between me and the dragons, and I have to say, these were the most menacing living creatures I have ever encountered.

Dylan had rightly guided me to this place to conquer my fears. The guides told me stories of humans coming to untimely ends—both adults and children being devoured because they were in the wrong place at the wrong time. My adrenaline was pumping the whole time I was there. A most magical moment occurred, however, that I share in my concerts before I sing the song "Nature Boy." And this moment remains imprinted in my heart as one of the truly amazing gifts I received from Dylan.

As I got off the boat on the second island—the biggest island, called Komodo— and was walking toward the land on the dock, I saw something dark and bent down to pick it up. I could not believe what I held in my hand. I was looking at a mama seahorse. I thought back to the time when I was engaged at the Opera de Marseille in France when Dylan was four years old. He had come running to me with a baby seahorse in his hand. He was so excited by that treasure he had found, and I was so astonished. I told him how my grandma and grandpa always had a little seahorse in their car for good luck. Seahorses have been a good luck charm in my life from the time I was a child. I looked at the perfectly formed mama seahorse in the palm of my hand as I stood on the pier entering the big island of Komodo, and exclaimed, "Dylan, you brought me to a far corner of the earth to let me know you are always with me, giving me such a sign!" I shook my head. Peace filled me, and I became unafraid of the finality of death as I felt him near. Love is eternal!

Revelation

A mama seahorse I found on Komodo Island and a baby seahorse Dylan found in France: mama and baby together again. (Photo by Jeralyn Glass)

"Son, you move me beyond words," I said.

He laughed. I laughed. Another miracle! I felt him right next to me.

As I viewed the big dragons on this island and heard more scary stories, I noticed my fear had shifted. As we motored back to Flores, I lay in the boat, watching the sky. Beautiful white cumulus clouds floated by. A huge rainbow appeared. I was in another state of being: safely and tenderly held in a stillness that was exquisitely beautiful.

All is well, I thought; there is healing happening in real time. I took a risk jumping into the unknown, and once again the net had appeared.

As I returned to my little hotel room in Flores, I tenderly put the seahorse I had found on a table and took a photo of it. It was indeed double the size of the one Dylan had found for me as a little boy in Marseille. A mama had been found in Komodo; a baby, in France. And a mama and baby were together across dimensions of time and space.

From Indonesia I flew to Thailand and the island of Koh Samui to continue my healing journey and spend time with Kevin and his daughter. I dove into a specialized program for loss, working with a mentor and meditation guide, receiving acupuncture and other healing treatments. The evenings were filled with sacred mantra. Kevin introduced me to one of the owners of Kamalaya, the award-winning wellness sanctuary there, and we enjoyed a meal together, speaking as if we had known each other for years. At the end of our time together, she invited me to return as a guest practitioner and musician to offer concerts and sound-healing experiences for the guests. I could hear Dylan laughing as I received her invitation to return to Thailand in a professional capacity. "Mom, what did I tell you? We have lots to do. It's only the beginning." The following year I began a fulfilling collaboration with Kamalaya that has lasted for years. I have treasured my times there with both the guests and staff, building the inner circuitry of a healer, honing my techniques and knowledge, and integrating sound and energy medicine, embodiment, and natural healing into practice.

And so, I continued walking a path that included gathering all the broken pieces and integrating them into a new experience of wholeness, and sharing them with others all around the world, guided by Dylan with the words of the Dalai Lama close to my heart: "Today is the right day to love, believe, do, and mostly, to live."

CHAPTER 9

INTEGRITY

To Serve at the Highest Level

Humans are a musical species and our brains co-evolved with music. This created specialized neural structures that respond to music. Music activates the entire brain in a way that nothing else that we know of does.

— DR. DANIEL J. LEVITIN

The first year after Dylan passed was a monumental time of new experiences. I longed for the familiar, yet I knew I had to step out of my comfort zone, trust, and dance into the unknown. I found that the more I explored sound healing with the crystal singing bowls, the better I felt, as did all the people I was sharing the sounds with. I had begun my tentative steps into my future by giving small sound baths regularly at my home and at my mother's home, and often she would invite medical doctors to attend. Everyone had strong positive experiences. One had complete relief of his low back pain; another's knee pain was alleviated. A woman who had lost her husband found peace, while others slept better, felt energized, or received an answer to a question they had been contemplating.

People loved these sound baths, and in the process of giving them, I was growing, building new neurocircuitry, and deepening my knowledge of the crystal bowls and how to play them most effectively. And they were powerful medicine for me! Their pure harmonics were able to bring the mind to a pristine quiet, a keen focus—different from other instruments. My grief had been dissipating, healing without medication, and it had been noticeable to all who knew me. As I transformed my loss and experienced a newfound sense of joy, I wanted to share with others what was possible through vibration as medicine. I wanted to provide a different kind of balm for pain, emotional instability, and physical ailments, serving through sound. Much to my surprise, nonmeditators who could not still the chattering of their minds found they could meditate easily during my sound baths!

However, despite feeling as though I was moving forward with the use of sound medicine and the crystal singing bowls, there were still many days when my heart was so heavy with grief, I could not get out of bed. I had to be patient with myself when the feelings became overwhelming. Healing requires a whole tapestry of initiatives. I was tuning in to what felt right for me, going into the pain through the "front door" and becoming my own guide through the uncharted territory.

One of my favorite authors is Rainer Maria Rilke, whom I read in both his original German and in English. His words accompanied me throughout my career and became especially poignant in the time following Dylan's death.

As Rainer Maria Rilke wrote in his *Letters to a Young Poet*:

> . . . be patient toward all that is unsolved in your heart and try to love the questions themselves, like locked rooms and like books that are now written in a very foreign tongue. Do not now seek the answers, which cannot be given you because you would not be able to live them. And the point is, to live everything. Live the questions now. Perhaps you will then gradually, without noticing it, live along some distant day into the answer.

I needed to be patient with all that was unsolved in my heart to eventually come to the answers, remembering that love is the most important ingredient in that alchemical process.

One day, immersed in my grief, I gave myself a serious talking-to. It was difficult to care about anything, and I certainly could not yet sing or teach. I heard a small voice inside me speak, and I listened. In response, I summoned the courage to call the Cancer Support Community in my city, an arm of a nationwide organization offering alternative healing modalities for cancer patients and their families. The largest professionally led non-profit network of cancer support worldwide, the Cancer Support Community has an important mission as stated by the South Bay Branch: "to ensure that all people impacted by cancer are empowered by knowledge, strengthened by action, and sustained by community. So that no one faces cancer alone." Thousands of people participate in the 200-plus programs this local organization offers every year.

I spoke with the director of programming, and I offered my services. I somehow knew that if I did not take my attention off how much I missed Dylan, I would drown in my pain. Every day I continued playing the singing bowls, creating spoken and sung affirmations, and exploring sound. In 2016, I began giving classes for cancer patients, sharing crystal alchemy sound-bath meditations to help reduce stress and pain. I prepared meditation scripts with different themes. At one point, though, I acknowledged it was time to trust. I let the scripts go, allowing the meditations to flow through me, supported by the structure I had created.

Today at the Sacred Science of Sound, we work with cancer, hospice, Alzheimer's patients, and veterans, as well as children and mothers in prenatal and post-natal care. We work in yoga studios and at addiction centers, we support people with grief, with physical pain, and mental health issues, and we participate in animal therapy programs.

I have personally been working with cancer patients for over eight years now and a number of different experiences have really touched me. One involved a woman who had stage 4 liver cancer. After a sound bath, she shared with me that she had never meditated with sound or music, although she had been pursuing a traditional meditation practice for 28 years. As she searched for the words to share her experience of the sound bath, in which I had integrated singing an inspirational song, she shook her head and said, "All I can say is that I am no longer afraid of death, as I have heard the sounds of heaven." Three different women patients who had each had a mastectomy shared, on three different occasions, that they had an unexpected miracle happen during the sacred sound bath. Each woman said, in her own words, that her surgically removed breast had energetically returned! The consensus was astonishment and the incredulous feeling that they were no longer missing a part of themselves; they experienced being whole again. One kept saying over and over, "I cannot believe what it feels like. It's a miracle! Thank you!"

I experience time and again how sound vibration used with intent can activate the transmutation and transformation of blocked, stagnant, and even long-buried energies to heal what was thought unresolvable. Sound as medicine is a gift for our mind, body, and spirit.

HEALING WITH CRYSTALLINE SOUND

Dr. Mitchell L. Gaynor, a pioneer in integrative oncology, was one of the first medical doctors to integrate the crystal singing bowls with his work with patients, beginning in the early 1990s. He also used Tibetan bowls, meditation practices, and patients toning using their own voice in his clinic in New York. He shares his revelations and techniques in his books *The Healing Power of Sound, Sounds of Healing,* and *Healing Essence.* Dr. Gaynor lost his mother to breast cancer at the young age of nine, which deeply influenced his career path. "I saw from my mother how people could have equanimity and inner peace in the presence of significant physical suffering and illness," he once said.[1] His dedication to helping people led him to integrate sound healing in his medical practice. In *The Healing Power of Sound*, he addresses the theme of tuning our human instrument and how sound readily connects us to our true nature and our ability to heal:[2]

Integrity

We're essentially like stringed instruments: One end of our wires is tuned to the infinite—our essence; the other end to the finite—the material world, our bodies, our egos. It's not that the infinite is better, and the finite is worse. If we are in tune only with the finite, we will be stuck in continual despair, frustration, and disease. If we are in tune only with the infinite, we may lose our ability to effectively negotiate our survival in the real world. Our goal should be to bring the infinite into the finite. Doing so enables us to exist in the present without being imprisoned by our wounds or egos, or the wounds or egos of others. Bringing the finite into the infinite, we'll never be undone by those who trigger past hurts with their insensitive words or actions. It's our birthright to be tuned to the infinite, and being so frees us from our perceptual prisons of victimhood, depression, obsessiveness, and chronic ill health.

When we utilize singing bowls or practice toning, the fine, harmonious vibration instantly entrains us to the frequency of our own essence. If we are open to the overtones and their resonance, we are reminded in a flash that our authentic selves are indivisible from "something larger"—the higher Self or God, however we define the Absolute. Our egos are tiny little patches of psychological terrain compared with the boundless spaciousness of our essence. The bowls therefore serve as a reminder that all we need to learn in order to heal is already within; sound vibrations are tuning forks that synchronize us to our own true nature.

Crystal singing bowls are being integrated into countless contexts today: graduation ceremonies, hospice programs, recovery centers, health-and-wellness retreats, and in preschools, middle schools, high schools, and universities. The bowls are being used in hospitals and private medical practices, in churches, and in yoga. People use the bowls with Reiki and as meditation tools, and they have become an important modality to support seniors;

veterans; women during pregnancy, birth, and postpartum; and even to support animals. They are being integrated into contemporary music, in chant and devotional music, and in classical music. They have been heard in concert at the Hollywood Bowl, the Greek Theatre in Los Angeles, and at the Metropolitan Opera House in New York. They are appearing in some of the most unlikely places, including weddings, baby showers, baptisms, sobriety celebrations, and sacred ceremonies. They are changing the way we experience music and transforming our social and personal environment.

I was very moved when two sisters traveled from abroad with their mother for a session. She had developed Parkinson's disease, and they were curious to see if crystal sound therapy could help her. I worked with a hand-held bowl she chose, a Rose Quartz Platinum alchemy heart-note bowl. I had her tone with me and entrain her voice with the bowl sounds. I moved the bowl around her body and especially played it adjacent to her heart. She started to cry and within a few minutes, her shaking began to lessen and then, miraculously, it disappeared altogether. It was incredible to experience her joy and her daughters' joy as she became calm and still. She purchased the bowl and took it home with her with instructions for how to use it in her daily routine of self care.

As I traveled on my own healing journey in the dynamic world of crystalline sound, so much continued unfolding for me. I knew what it meant to study an instrument and fine-tune my technical skills. Yet this was a different kind of learning and practicing: the crystal bowls are experiential, creative instruments, and they were teaching me many subtleties through their unusual harmonics in combination with the theory of music and the science of sound. The healing experiences ignited by the bowls were as varied as the bowls themselves. I wanted to share everything I was discovering with others, so I created a four-day immersive training diving deep into the world of crystal alchemy sound healing.

Today, the Sacred Science of Sound Crystal Alchemy Trainings are one of the leading training programs in the world, offering more than 120 hours of education over four levels. Our student body comes from different European countries, England, Australia, Malaysia, Japan, Hong Kong, South America, South Africa, Canada, and Mexico, as well as locations throughout the United States. We explore in depth each person's individual alchemy, using creative exercises that help unlock and free their human instrument; we look at music theory, bio-energetics, and the science of sound, and when our students complete the program, their graduation includes a public sound bath.

Our graduates are implementing sacred sound into their daily lives and the lives of countless people around the world, and they, like me, are gratified and fulfilled by the powerful transformations they see taking place as a result of their work—from the child who was relieved of her panic attacks and the boy with autism who shifted dramatically, to the middle-aged man who found lasting relief from his chronic back pain, to the woman who finally conceived, and so many more! The students are taught how to work with clients in person and virtually, including how to ensure the highest quality of sound possible with today's technology in an online setting.

> ### Sound Rx
>
> This experience, shared by Susan Eva, a BioEnergetic Synchronization Technique (B.E.S.T.) practitioner, Energy Codes master trainer, and an advanced Graduate of Sacred Science of Sound, illustrates the healing power of the bowls:
>
> "I have been working for two years with a woman I will call Sophia. She had been diagnosed with a glioblastoma, which is a fast-growing, aggressive tumor in the brain. Western medicine treatment protocols usually include chemotherapy or radiation, which are used to slow the progression and reduce symptoms. There is no cure and life expectancy can vary, yet it normally is between 12 and 18 months from diagnosis.
>
> "Sophia works in the medical profession, and she began regular chemotherapy treatments upon diagnosis. A relative referred her to me, and we began working together right away using a combination of two modalities for the energy of the body. The first was the B.E.S.T. The second modality was the crystal singing bowls. The crystal bowls bring

in the vibrational frequency of nature and help bring the body back to our natural state of well-being. I used a combination of four different bowls in the key of B major, all tuned to 432 Hz, which is the frequency of nature and healing . . .

"Using these beautiful bowls together not only relaxes the body, but it also puts the body into a state of harmony and balance from a vibrational perspective, which allows the body to utilize its innate power to heal.

"Due to the restrictions that were in place at the time, our sessions together were not in person. We met regularly via computer using Zoom, and each session included a B.E.S.T. treatment and a 30-minute sound bath. In many instances I was playing the bowls live, yet while I was out of the country, I used a variety of recordings of the bowls that I had created specifically for Sophia.

"Between sessions, Sophia would use the practices of the Energy Codes that were also part of Dr. Sue's teachings. During this time, she was able to maintain a regular exercise regimen and had the energy to navigate her life normally. She continued the chemotherapy protocols and had periodic MRIs to monitor her progress. At each MRI scan, the tumor was smaller, and her doctors commented that she should not be feeling so good or having the energy to do all that she was doing.

"After about a year, Sophia's most recent MRI results indicated no evidence of the tumor. Her doctors were hesitant to acknowledge the tumor was gone, as they are not accustomed to seeing these tumors dissolve. Yet they advised her to keep doing whatever she was doing.

"As Sophia is now approaching the end of another year, with continued MRI reports of no tumor being present, her doctors have advised that just one more clear MRI and she will be able to reinstate her driver's license and resume driving. Her color is good, her energy is vibrant, and she lives in gratitude daily for the gift of regaining her life back. She is fully experiencing the joy of life with her husband, her children, her grandchildren, and her extended family and friends. We met in person, and I was inspired even more by this woman who made a choice to seek other options and set her intention to healing completely. She is a testimonial to infinite possibility and the value of asking herself, "What if?" She recently saw the oncologist and told her she was nervous about the tumor coming back. The oncologist looked her straight in the eyes and said, 'You are cancer-free. Go live your life!'"

GIVING BACK AND LOOKING FORWARD

We know now that everything is energy and sound vibration, including us, and that sound can affect matter, thus creating physical changes. Some exciting research and discoveries from Stanford University are opening phenomenal possibilities showing us that "acoustics can create new heart tissue." Frequency and amplitude put cells in motion, guiding them to a new position and holding them in place. Sound can create new tissue to replace parts of a damaged heart."[1] The possibility of the regeneration of cell tissue broadens the definition of what it means to "tune the heart" with sound, including replacing damaged tissue.

My vision is that every human being will have access to healing frequencies and music medicine for their health and wellness. I am committed to continue expanding the work of the Sacred Science of Sound, making healthier, more empowered, joyful, and fulfilling lives available to everyone through natural healing interventions that include energy, sound, frequency, and music. In June 2023, the Laboratory for the Science of Music, Health, and Wellness at Minerva University was formed and began its first program with seven international fellows. I had the pleasure to introduce them all to the crystal singing bowls. One of them, a young man from Canada, is a musician—an accomplished pianist—and he was astonished at what he felt in the short time he played them; he didn't want to stop. He said he had never heard any sounds like these. I watched his eyes light up as he and the other fellows experienced states of deep relaxation and mental clarity. Minerva has a presence in seven different major cities—Seoul, Hyderabad, Berlin, Buenos Aires, London, Taipei, and San Francisco—and the students felt it would be a good idea to place a meditation set of three bowls in all Minerva residences. They were excited to note that this is "sound health" and all students everywhere should have access to it.

I am excited that the lab will include studies employing scientific research to explain how the crystal singing bowls create the powerful results we are seeing. We are collaborating with the scientists and fellows in designing and implementing the experiments and the research that will do so. There are no known side effects to participating in a sound bath with the alchemy singing bowls and the positive impact they are having on people is remarkable. Even if you are only seeking to receive a greater feeling of calmness and relaxation, already that's a huge win in today's challenging world. As we know, stress alone can cause many health problems. And the bowls played with integrity and intention are powerful stress-busters indeed.

CHAPTER 10

CURIOSITY

Common Questions about Singing Bowls

The most beautiful thing we can experience is the mysterious. It is the source of all true art and science.

— ALBERT EINSTEIN

Sounding the [perfect] fifth is a general sound tonic. Some of the benefits of the fifth are: alleviates depression, enhances joint mobility, balances Earth with spirit, and directly stimulates Nitric Oxide release.

— DR. JOHN BEAULIEU, "A BEGINNER'S GUIDE TO SOUND HEALING WITH TUNING FORKS"

The room was full, the candles were lit, and a feeling of tranquility permeated the space. It was February 14, 2019, and I was performing at a big studio in Los Angeles, leading my third World Sound Healing Day event. Jonathan Goldman initiated this celebration in 2003 (worldsoundhealingday.org) to encourage people around the world to create healing sounds encoded with the intentions of Love, Kindness, and Compassion. His goal was, and still is, to help raise consciousness so that together, we can manifest Global Harmonization. World Sound Healing Day invites everyone with the heart and mind to make a difference through sound vibration—musicians, sound healers, teachers, and leaders all around the world—to join in with sound offerings of their own.

My plan for this particular evening was to lead a meditation with the theme of releasing generational patterns.

I had chosen my bowls based on the three structurally stable, sacred intervals of music: the octave, the perfect fifth, and the perfect fourth. Intervals are the distances between two notes, and when implemented properly, create a powerful musical structure for healing. (For more on intervals, please turn to the glossary.) I had integrated a number of these stabilizing musical intervals in combination with specific alchemies, including Saint Germain, which is said to be the custodian of the Violet Flame of Healing. The energetic theme of the curated set supported embodiment, cleansing, releasing, and the opening of the heart, encouraging love, humility, and forgiveness. The set was tuned to 440 Hz, the same as modern-day music, which made it easily accessible to all.

Most people have now heard of a "sound bath" and understand it is an experience of deep relaxation with many possible additional health benefits. You don't need to undress or put on your swimsuit; you don't need a bar of soap, and there is no water involved! But what exactly is it, and what makes an effective one? People are becoming more curious about singing bowls, and I often get asked these questions:

- Are your sound baths and sound-healing concerts planned and rehearsed?
- How do you know which bowl to play and when?
- Do you have a structure for your sound baths?
- Are they always the same, or do they vary according to whether you're playing for a single person, a small or large group, a special event?

The answers: Some planning and rehearsal go into a sound bath, and yet the experience happens in the moment. The musical structure is always in the foreground, creating a container of safety, stability, and trust. Often, I begin with the lowest note in the set and slowly expand upward from there. Implementing techniques for grounding, setting an intention, and knowing how to facilitate sacred space are vital. I reflect on the purpose of the sound bath, set a theme for the experience, determine the length of time it should ideally last, and choose an appropriate set of bowls. They will always vary according to the group and the event.

The questions I ask myself include:

- Which musical key will I play, and why?
- Will I use a pentatonic scale set? Will I use a chakra set? Why?
- What tuning am I using: 432 Hz, 440 Hz, 528 Hz—or a mixture?
- What musical intervals will I use for structure?
- Am I adding a beat frequency (the gentle, shimmering sound)?
- Will I share a guided meditation? And if so, what will it include?
- Am I singing a song? Adding toning, mantra, or light language (healing sounds expressed through certain vowels, consonants, rhythms, and syllables to activate, awaken, and elevate)
- How will I start, and how will I end? Will I use a favorite quotation, read a poem, or say a prayer?

The first album I created after Dylan's passing, called *Forever Love*, was scored for singing bowls and included classic American songs by the arranger, composer, and orchestrator for film and television Perry La Marca. Musical key changes are integrated into some of the songs and, in a few of them, when I'm performing, I need to have extra bowl notes for the modulations into two different keys. This is where musical knowledge brings a wider possibility of expression with the bowls and why I include some study of basic music theory in my Sacred Science of Sound Crystal Alchemy Trainings.

However, in a relaxing sound bath, using the swirling technique, gentle tapping, and perhaps a guided meditation, this theoretical knowledge and musical skill are not necessary. It is important, though, that the bowls are tuned to one another and play harmonically, and one should know one's set, including how the notes relate to one another and the energetic qualities of the alchemies. People have often shared that they did not like certain sound baths they had attended and wondered why some sound experiences leave them with negative feelings. Although anyone can learn to play the singing bowls, there are certain rules of harmonics to be followed, and a discordant sound bath can bring discomfort that lasts for days.

I began the Generational Healing Meditation I played for the 2019 World Sound Healing Day by grounding and centering. I stabilized the energy in the room and created safety. I intoned an invocation and set an intention, then focused the attendees' attention on their mother's side of the family, inviting them to release patterns that were no longer serving them.

"Increasing your awareness," I said. "Breathing deeply in your belly. Letting go. Limiting beliefs and patterns of behavior. . . Feel. Release. . . your mother, and her mother, and her mother's mother. . . repression, anger, withholding, powerlessness, abuse: they are no longer yours to carry." Sighs. Sobs. Deep breaths. Stillness. Rebirth.

The three octaves of C notes I was playing bridge Heaven and Earth as a universal, vibrating "string," each one doubling the previous one's frequency as they ascend from the lowest C to middle C and finally to high C. This musical structure connects the human being with both the earthly and the celestial. (This is a crystal bowl example of Dr. Gaynor's comments on page 107.) I love to have two or three octaves in every set for exactly this reason. They deepen the sound-healing experience by simultaneously supporting embodiment and expansion.

After we cleared the maternal lineage, I played the bowls alone for 15 minutes. Sometimes it is important in the structure of a sound bath to leave space with no vocal guidance so people can land in the pure crystalline vibrations and feel. We then moved to the father's lineage. I began to lead the spoken meditation into the release of generational patterns on the father's side that no longer served: "Control, aggression, abandonment; your father, and his father, and his father's Father. . . repression, anger, withholding, punishment, abuse: they are no longer yours to carry. . . "

While we were focused on the father's side, the big Saint Germain grounded root-chakra bowl I was playing started to screech with a high overtone. It surprised me, and I adjusted my technique. I had been playing that bowl for a few years, and it had never made that sound before. Just 20 minutes earlier it had resounded its usual round, lush, deep-octave timbre. It took me completely by surprise. But as I swirled around the rim of the bowl, the high-pitched screeching continued. Everyone was deep in the sound immersion and showed no sign of disturbance. In that moment, I recognized the bowl was bringing a kind of penetrating frequency, helping the group cut through energetic density and release patterns being held on the father's side individually and collectively. The intense sound was piercing through and beyond our group. I understood. This had not happened on the mother's side or at any other time during the sound bath. The tone was quite shrill for me as I was so close to it, and it went on for some minutes. I was waiting for someone to get up or express discomfort, but no one did, and when the sound bath was over, the energy in the room had completely shifted.

Afterward, people shared their experiences of tremendous release, of long-forgotten, tucked-away remembrances that had melted, and of their feelings of heaviness entraining and then releasing with the bowl sounds. Some shared how physical and emotional pains disappeared, replaced by a lightness of being that then easily anchored in them. Some of the men, especially, expressed an unusual sensation of relief from guilt and shame and a release of long-held anger.

This experience gave me a reference point for how the pure crystalline instruments, combined with our intention, become living instruments of love in action, especially working in tandem with a practitioner who uses both refined technical skills and the unique expression of their own soulful gifts to facilitate the highest good for all. It is remarkable how consistently this works, and I have learned to trust the bowls and myself as we collaborate with the attendees and create a container in which healing can occur. This is available to you as well, and whether you have a set of bowls or you explore the hundreds of free sound baths on our Crystal Cadence YouTube channel, I wish you many grace-filled moments as you continue to discover how sound can support and transform you.

A DEEPER LOOK AT WORKING WITH THE BOWLS

> Each celestial body, in fact each and every atom, produces a particular sound on account of its movement, its rhythm or vibration. All these sounds and vibrations form a universal harmony in which each element, while having its own function and character, contributes to the whole.
>
> — PYTHAGORAS (569–475 B.C.)

There are many ways of working with the bowls. For example, at wedding ceremonies, I normally choose to play in the key of F, the key of the heart—committed love in action. And I may choose to include as alchemies the couple's favorite stones or metals, such as Gold or Rose Quartz. As we created the Forever Love album, we chose to arrange "Somewhere over the Rainbow" in A major, the key of the third eye, to add the subtle intention of strengthening vision and intuition. The bowls I played included Apophyllite and Azeztulite, which activate Reiki energies and higher realms of consciousness. The possibilities are endless.

Some of the technical skills required to play the bowls include:

- managing the mallet or wand
- knowing how much pressure to apply
- knowing when to stop swirling the mallet to allow the bowl to ring and sing
- finding the sweet spot on each bowl and tapping or gliding it with just the right amount of effort
- knowing how often to change tones

These skills are required whether you play the bowls by swirling the mallet around the rim of the bowl or use the "bowing" method, in which the mallet is used as if bowing a stringed instrument. With this method, you do not play around the rim of the bowl but rather around its middle. This involves a different flexibility of the wrist.

When I teach people how to play the bowls, I like to begin by holding the wand like a pencil. A gentle swirl once or twice around the rim, and the bowl begins to sing. Easily and effortlessly, the crystal vibrations resound. I like to sweetly tap the bowls with the wand like a gentle kiss, never too forceful. These tender taps can nudge and awaken us, while the swirling can help us drop inside. Every person has a unique style of playing. Clockwise motion brings energy into the body, while counterclockwise is releasing in nature. There are different kinds of mallets we can use to play the bowls. One is made of silicon or a combination of silicon and glass made in Germany, and it eliminates the friction sound some people object to when you play the rim. The other mallet is made of suede, which is the preferred mallet of many music producers and musicians, including myself, as it amplifies vibrant, natural tones which are clearly the sounds from an acoustic instrument and not a synthesizer. I normally use both wands when I play, to get different kinds of sounds. I still am in awe over a miraculous communication from Dylan, and how an additional piece of important technical information was relayed to me through one of my clients, Angela Weisman, in

Dayton, Ohio. One of the women in her meditation group, Stephanie Hamiel, shared an experience which clarified any questions about the sounds the different wands produce.

With gratitude, Stephanie shared that "The first time I remember being able to reach my safe place in the void was during a woman's gathering. My dear friend Angela played the bowls and spoke softly about the story behind them. As she was playing the bowls, a beautiful, bright light, with a blueish hue, came through. I don't think I can ever truly convey the beauty of it, unless you had experienced it yourself. I looked at my friend and asked her 'Who is Dylan?' Angela was shocked. The name came from this Light. It emitted a protective, masculine energy. Angela replied, 'Dylan is Jeralyn's son.'"

Stephanie continued, "Dylan made his presence known to give me reassurance that I was safe here. Angela graced the group with her gift of a serenely powerful meditation. I remember crying and releasing a massive amount of guilt I didn't even know I was carrying. It was intense. I felt so protected in this space, that I was able to finally reach 'home.' When the session was wrapping up, I remember someone in the group asking, 'Do the bowls have to make that scraping sound?'"

She was referring to the scratching sound the suede mallet makes. Without hesitation, Stephanie replied, "Yes, it is necessary for removing layers of negative, trapped energies that may have attached to us like barnacles. It helps scrape them away and clear the pathway for Light."

"In that moment I knew from Dylan's incredible blue Light that this statement was absolutely true, and no one could persuade me otherwise." The unforeseen revealing of sound wisdom from Ohio to California exemplifies the mystery of the sublime, the gift that saying yes to the bridge between Heaven and Earth can deliver.

In addition to the wand techniques and just as important (if not more important) than playing the bowls with a focus on technical mastery is to play with Authentic Expression. This second "technique" involves a completely different skill set that nurtures the development of one's unique gifts, talents, and interests, or "personal alchemy." The cultivating and strengthening of one's personal vibrational signature grounds trust, confidence, and the release of any limiting beliefs, including disempowering energies that might block one's natural expression. This technique is creative, expansive, and nurturing; it helps remove doubts and any feelings of unworthiness or not being good enough. It encourages true presence and entrainment with the bowls, which then activate and amplify the resonance of pure Spirit.

What emerges is the free and easy flow of vibrant energy. The person playing is nourished, as are the receivers, and that beautiful energy is amplified through the singing bowls. Authentic Expression is at the core of everything I am teaching, as it encourages generosity of spirit in combination with sacred sound. The bigger intention of service through sacred sound is always at the essence of the Authentic Expression technique.

CRYSTAL SINGING BOWLS AS A MUSICAL INSTRUMENT

What is the secret ingredient that makes a sound-healing event memorable, moving, and transformative? I'd say it's a sound healer who knows their bowls—who knows their alchemies, properties, tunings, and the musical key of their set. They also understand which notes make up the sacred intervals of the perfect fourth, fifth, and octave, and recognize the importance of musical structure. An effective sound healer knows the purpose of the event at which they're playing and leads with intention and presence.

When I am using the bowls as a true instrument and recording or playing live with other musicians, it is important to clarify musical keys, tuning, and a general shape and structure for our song or meditation music. Within that musical structure, we are free to create in the moment. That grounding and clarity, in combination with our talents, fluidity, flexibility, and our presence, allow us to respond in real time to each other's playing. This combination gives rise to unpredictable and extraordinary beauty through sound. The tones of the bowls lend a very particular color to music that can help focus the mind, open us to the soul, release anxiety and depression, and bring a relaxed state of being simply and easily without any trying.

When I was playing live with singer-songwriter Jhené Aiko at a sound-healing performance in Los Angeles, the meditation and songs took us—and the audience—to an unexpected place of pure awareness, renewal, and rapture. We had rehearsed our structure, and then responded to each other in the present moment. It was a gift for us all, a sonic tonic for our senses. And it was glorious! This highlights the importance of attention to presence, openness, receiving creative impulses, and deep listening for both the musicians and the audience. So much is possible through music as a healing modality when we land inside ourselves and choose to express our true nature. We open to becoming a part of something bigger than ourselves, leaving our egos aside. This is the bliss that is possible with collaboration and aligned communication.

I am a huge advocate of this. In another such instance, I was invited by my friend, *New York Times* best-selling author and world-renowned transformational teacher Marci Shimoff, to create a potent sound healing experience for the 2023 Inner Circle Year of Miracles Retreat held online and broadcast to participants around the world. The key words given to amplify the theme *Illuminate* were joyful, reverent, engaged, sweet, and revelatory. I built a bowl orchestra tuned to 440 and 528 Hz, with exactly these energies and these intentions, created unique, stabilizing harmonics, and used particular alchemies such as Phenacite, Andara, Divine Kryon, Palladium, and Saint Germain, also adding an effective beat frequency made with two middle octave Supergrade bowls. What happened was astonishing, something I had never experienced before.

The bowls played in a heightened manner, almost otherworldly, as if they had an assignment beyond what I could possibly have known or understood. At one point in the first half of the 45-minute sound bath, a portal opened. I crossed a bridge into a place of complete expansiveness with a perpetual emanation of radiant, indescribable brilliant light constantly infusing in me. In this expansive, limitless state, I continued playing as words, sounds and tones emerged. I was elated and in rapture.

When it was finished, I stayed in a state of deep bliss for five full days. The infinite had become my new point of reference. My consciousness shifted in a miraculous way and the theme of *"Illuminate"* was indeed the energy grounded in that transcendent experience. Sound curated and structured from the heart, with integrity and knowledge, is a powerful transformative medicine.

CRYSTAL SINGING BOWLS IN A THERAPEUTIC SET

One of my most moving experiences using crystalline sound was with the first hospice patient I ever played for. His caretaker was in the home. I played at his bedside and used a Phenacite alchemy bowl. Phenacite is a high-frequency stone that is said to activate light. It is one of my favorite alchemies to work with, and I find it supports a connection to higher realms.

Exquisite sounds floated through the room, and the gentleman I was playing for drifted in and out of physical presence. It was a privilege to witness his intimate farewell to this life experience as his soul took flight and his body, which was very ill, came to know peace.

In the traditional type of sound bath that I create and guide for cancer patients and their families, the bowls are a harmonic set, and my intention is to land everyone in a place of deep relaxation and renewal, where the mind has ceased its chatter and people drop easily into inner peace and trust and allow any fears or anxieties to release. From this grounded, centered, and open space, miracles happen. I will always remember a man who accompanied his wife to a sound-healing circle; after the sound bath, his heart was wide open. He vulnerably shared how much he had looked forward to the time when both he and his wife retired. He had never dreamed she would get cancer. He cried. He expressed an acceptance to go through whatever would be necessary and shared that the sound-bath experience had brought him strength and fortitude.

TRANSFORMING TRAUMA WITH CRYSTALLINE SOUND

When I play for veterans or seniors, I like to begin with a short explanation of healing vibrations so that audience members understand a part of the science and history of sound. I follow this by teaching some breathing techniques; then, depending on the audience and the theme of the event, perhaps I will sing a song for them such as "No One Is Alone," by Stephen Sondheim, from the *Forever Love* album. Then I will let that flow into bowls playing solo, leading into a guided meditation, and then close with a prayer or another song, such as "Count Your Blessings," part of *A Gratitude Meditation*. If I am playing for a company, I create a different structure based on their program request and employees' needs.

I'll never forget what happened at a formal event for veterans; a female sergeant—who had just experienced the bowls for the first time—jumped up after the presentation/meditation and exclaimed, "I have been given every kind of medication to help me with my anxiety and pain, and I've never felt anything like I just felt. Relief! Nothing, and none of these endless medications, has ever quieted my mind as this just did! I'm stunned! I've been a clinical test animal. I've been raped; I've seen death and fought in war; I've lost close friends. I never thought I could feel this way again! I'm in bliss and feel deep peace. Dear God, this is incredible. Bless you!"

> ### Sound Rx
>
> Sara Bayles of Soul Wave Wellness is an advanced graduate of our trainings and a well-loved practitioner in Los Angeles. She has created a unique curriculum and program for veterans and a specialized program for horses.
>
> "I've played for a blind veterans' program at the Long Beach VA hospital over Zoom and in person, and it has always been an amazing experience. But one time in particular stands out. A man asked me, after I played for a small group for about 45 minutes, if I could

help him understand what sensations he'd experienced in his heart as I'd played. I explained that sometimes our bodies hold emotions and that perhaps he'd had a release of energy in response to the music. I played my Salt F bowl for him on its own. I explained how each bowl corresponds to a different area of your body or energy center and that my F bowl corresponded to his heart chakra. When I played it for him by itself, he immediately responded enthusiastically, saying, "Yes, that's it! Would you mind if I took this bowl home with me?" I was so happy to hear this from him. And, of course, he was kidding.

"I have what's called a 'chakra set,' meaning each bowl corresponds to a different energy center or chakra. If I could put the qualities of my set into a sentence, it would be that they clear the way with love. To me, this means that the alchemies of my bowls have many clearing qualities—like salt for detoxification. And many of them also have elements of the heart or unconditional love. I strive to be of service with my playing, to be a channel for the highest good with my bowls, and the conduit that allows them physical expression in the world, where they can work their magic and heal."

Nada Hogan is another of our Sacred Science of Sound practitioners helping veterans, and she uses the singing bowls in combination with bio-energetic treatments and acupuncture in her sessions. She shared with me a remarkable story about a Vietnam vet suffering from excruciating headaches. "Al was unable to close his eyes and fully relax when he was on the treatment table for acupuncture. I decided to incorporate my singing bowls, and in five sessions, he was able to trust the work and fully relax and close his eyes. Trust for a veteran with PTSD does not come easy, and especially not for Al. He had tremendous pressure in his head, and during the first session, the pressure and headache disappeared! His headaches never came back. The pressure in his head, however, did return; yet after every session with the bowls, the pressure is relieved."

"He told me, 'Nada, no one else even tried to understand the root cause of my discomfort. The bowls are, miraculously, the only thing that has worked. I would still be suffering without the bowls.'"

Sometimes with sound alone, we enter a state of amplified awareness that allows us to remember life experiences that were stored deeply in the subconscious. Sound can help us access memories and integrate them consciously, allowing us to have a bigger perspective of our own lives and facilitating a return to wholeness. This is exactly what happened with Lois during her first sound bath.

> The sound bath was a new and very powerful experience for me. I went unexpectedly deep, to a place I can only describe as "Cosmic." I felt myself lying comfortably, being cradled, and held in a nurturing embrace. Suddenly, I had a flashback from a traumatic time in my childhood that I have no real memory of. In an instant, it was clear: I remembered being on a stretcher as they loaded me into an ambulance. Afterward, I awoke in a hospital room as I came out of a coma. I was calm; I had no panic and no fear. It was astonishing how the sound vibration took me almost immediately to that time and place as a young child when I was in a coma for 10 days! For the first time in my life, I was reliving what happened to me at the age of seven when I checked out. The sacred vibrations took me to an incredible place of peace, love, and calm where I could be curious. Revisiting this time in my life in a safe manner allowed me to feel that I am always taken care of and that I am held. Being able to witness that event without reliving trauma or physical pain gave me a higher perspective of life. The sound bath made me recognize there is a place beyond this physical world that is filled with beauty, love, and peace.

For these gifts of relief and healing for people all around the world, I am grateful. It is challenging to suffer and feel alone in that pain. Sound medicine is bringing comfort and relief to many. It is wonderful to be a part of this advancement and witness the growth. Humanity is evolving to receive and integrate intentional sound and music as regular tools for well-being.

Sacred vibrations show us a new way to perceive, teaching us to turn our focus inward and allowing us to experience expansion on many levels. In this flow state, our true essence can express itself with grace and ease. When the player and the crystal bowls are attuned, the effects of the harmonic resonances are transformative beyond words.

As we continue exploring this fascinating world of sound with great curiosity, we find joy and new ways to tap into our infinite potential. May the flow of sacred sound be with you.

CHAPTER 11

GENEROSITY

Connections to Source

There is no greater reward than working from your heart and making a difference in the world.

— **CARLOS SANTANA**

Music is a language that calls on us to be generous of spirit and focused in our purpose. The more I explore creating music and sound-healing experiences with the alchemy singing bowls, the more I see how these unique vibrations inspire the heart to open and the spirit to soar. As a practitioner or musical artist, the possibilities are profound, miraculous, mystical, and transcendent. And generosity is an integral component in the energetic mix. A conversation with my dear friend Dr. Sole Carbone highlighted for me many aspects of the singing bowls' pure quartz crystalline sound and its place in our lives. Dr. Carbone has more than 20 years of experience in holistic practices and more than 30 years of experience with sound. She is a sound-practitioner graduate of the Sacred Science of Sound Crystal Alchemy training program, and she shared one of her healing experiences in a Sound Rx box in Chapter 5.

I asked her about the relationship between nonordinary states of consciousness and the role sound can play in the healing process, especially for trauma and deep wounding. Sole's firsthand experience began at age 14, when she fully immersed herself in the Vedic tradition of Bhakti yoga, a path of devotional service and daily mantra chanting. At the same time, growing up in South America, she was deeply drawn to learn and practice the shamanic traditions, specifically those of the Q'ero people and, later on, holotropic breathwork. She has also integrated crystalline sound into her practice. I was curious to hear her perspective on the mechanism through which sound vibration and music integrate mind, body, emotions, and spirit.

She tells us, "First, I would like to reflect upon the stepping stone of a lifetime in sound journey—the fact that during my darkest, loneliest times, chanting mantra saved my life and changed me forever. . . If there is one element that can consistently be traced throughout human history and evolution, that would be sound. Human beings have used sound as a medium not only to communicate with the outside world, but also to communicate with our inner world and even with the world of the unseen, the subconscious."

She noted that if we were to draw an analogy between ancient healing practices and modern, cutting-edge, therapeutic settings, we would find some astonishing similarities. As much of a dichotomy as it may seem, the figure of a therapist in these settings could perfectly well be compared to that of a shaman, a healer, a seer, and a counselor. Dr. Carbone's ancestral medicine people understood health and well-being as a constant interplay of inner and outer forces in rhythmic balance. Those few who were responsible for the well-being of their communities seemed to have the capacity to bridge the worlds, to see beyond what was evident, and to find there the root cause of what was presenting itself as a problem. They did not make a distinction between physical, mental, societal, or environmental problems; they understood that rhythm was the energy holding reality together and that the so-called "problems" were simply the result of a disharmony among the natural rhythms.

"What made our medicine ancestors stand out among the other members of their communities—even above rulers—was not their capacity to heal but their capacity to see beyond the seen, beyond what was evident," she continued. "With this expanded vision, they were able to guide others to the best possible solution."

Although this may seem like a romanticized view of primitive societies, what was once thought to be in the realm of the mystical is now being explored and understood by scientists, doctors, and therapists.

Dr. Carbone stressed that what gave our ancestors their "seeing" abilities was their capacity to access nonordinary states of consciousness, meaning that they were able to induce within themselves different wavelengths of brain activity that we now know to be responsible for these so-called altered states.

In many cases, she said, the healing ceremonies they relied upon involved the use of sound and psychoactive substances such as psychedelic mushrooms, roots, and plants. She describes further that during these ceremonies, all members involved would be induced into an altered state that would allow them to connect, to tap into Source, and to have epiphanies or breakthroughs. What this meant was that through these transcendental experiences, they would be able to perceive the problematic situation in a different way, to understand something about it that they were not able to comprehend while in an ordinary, wakened state. As a result of this new way of seeing, they would also have a new understanding of the situation and a new way to approach it—what is commonly known as a "bird's-eye view" and symbolized in the Andean traditions by the condor (or *kuntur*, in the Quechua language). Dr. Carbone went on:

> Since 1938, with the discovery of lysergic acid diethylamide, better known as LSD, there has been a gradual increase of interest in researching the possible therapeutic effects of psychedelics and altered states of awareness. The studies began in parallel in the 1960s at the Psychiatric Research Institute in Prague and at the Harvard Center

for Research in Personality, and involved Richard Alpert, better known as Ram Dass, and Dr. Stanislav Grof, one of the founding fathers of transpersonal psychology. Initially, the way in which the Western world approached the subject led to public condemnation and prohibition of psychedelic substances which, paradoxically, ignited more popular curiosity and a thirst for exploration.

On the one hand, more people, including doctors, scientists, and psychotherapists, began to travel to various indigenous communities around the world to have firsthand experiences with psychedelic substances and direct contact with medicine wisdom keepers. This led to the current understanding that, although the use of psychedelic compounds can be highly beneficial, it can also have detrimental consequences when it is decontextualized, misused, or unsupervised.

On the other hand, people such as Dr. Stanislav Grof began to explore the possibility of accessing these nonordinary states without the use of psychedelic substances. He turned instead to primordial sound and rhythm, the essential components of all ancient healing practices. Dr. Grof understood that by combining rhythmic tribal music and rhythmic breath, we could endogenously replicate the chemistry and positive effects of exogenous substance ingestion without detrimental side effects.

After more than six decades of experiences and research, these two very different approaches shed a new light on the consensual understanding of healing processes and of reality itself. Alongside these modalities, new scientific modalities such as quantum science, epigenetics, and neurobiology are beginning to validate the principles applied in ancient healing practices. They posit a new paradigm that challenges the consensual belief about reality and brings forward the ancient understanding of the nonmaterial roots of our material world, the interconnectedness of all realms of reality, and our relationship with them, thus inviting us to reconsider our innate capacities.

Generosity

In this new paradigm, understanding the rhythmic subconscious interchange that we continuously have through vibration, frequency, and sound is fundamental to further understanding and developing ourselves.

Dr. Carbone concluded with a profound concept for our times of rapid change: "Incorporating this modality can very well serve as a tool for the evolution of consciousness: a substance-free option in which shamanic practices and modern medicine become indistinguishable from each other."

This type of breakthrough or rewiring of our thought patterns can also happen in the most unexpected and lighthearted ways. I was privileged to play singing bowls for friend and well-known clothing designer Han Feng, both in Shanghai at her home and at her design studio in New York. Han Feng's exquisite style made her the one chosen to create breathtaking costumes for the late Anthony Minghella's production of *Madama Butterfly* at the Metropolitan Opera. She had invited her friends to her flat one rainy day while I was traveling in China with my mother, and asked me to play a sound-bath immersion for everyone. As I began playing, everyone closed their eyes and became very still, quickly entering into a state of deep relaxation. Suddenly, one of Han Feng's dogs jumped into my mother's lap and promptly fell asleep. He was in heaven, but my mother was not! My mom is a cat person, not a dog lover, and it was time for that to shift! As the dog bounded right into her lap, my mother looked at me with raised eyebrows as if to ask, "Jeralyn, what do I do?" The room was quiet; everyone was deeply immersed in the beauty of the sounds I was playing.

I looked back at her with a strong, nonverbal communication: "Mom, chill. Hold him tenderly. And please, receive the meditation!" Mom certainly had a rewiring of her mind-body and emotional system that day. She relaxed and received, and it was so touching to witness. Afterward we shared our experiences in the group and laughed at the sweetness of it all. Restructuring can be accompanied by joy!

We still chuckle about my mom's enlightening experience with an animal and sound vibration on that rainy day in Shanghai. It certainly helped shift her feelings about dogs! But can we understand what really happened with her? Crystalline sound brought her back to an innocence. She trusted, let go of a belief and reprogramed herself. Crystalline music guided her to return to Source. As Dr. Carbone explains, when we use crystal singing bowls in a therapeutic setting, we open a gateway of communication between the inner and outer worlds by temporarily disabling our subconscious survival mechanisms. The sound frequencies of the crystalline structure of the bowls can entrain with our brain waves, promoting expanded states of awareness, relaxing the body at a cellular level, and naturally inducing the necessary biochemistry to generate new neural pathways.

The Earth Is Healing with Crystalline Sound: The Land and Her Creatures!

The benefits of crystal sound healing are not limited to people. Holistic wellness expert Amy Bacheller—among others—has also used crystal bowls to good effect in land and animal healing, as well. Amy has taken all three levels of the Sacred Science of Sound Crystal Alchemy Trainings, and I curated several of her sets, including her Elemental Set of bowls, which is tuned to 432 Hz. Each of the alchemies in this 14-bowl set represents the earth's elements: Sunstone and Citrine for fire; Etched Fairies and Saint Germain Sky for air; Etched Dolphins and Ocean Gold for water; Sedona, Red Rock, and Mount Shasta Serpentine for earth; and Selenite and Divine Feng Shui for ether.

"I use this set for sound baths with nature-based themes such as connecting with plants, animals, or nature spirits, as well as for celebrating the cycles of the sun and moon," Amy says.

Here she shares one of her land-healing experiences: "In my holistic healing practice, one of my clients asked if I could do a land clearing and healing for her newly purchased property in Virginia near a Civil War battleground. Feeling the call of the bowls to assist with this request, I immediately said yes. We met virtually on Zoom, with me in California. Spanning the country from coast to coast, we discussed the history and challenges of her new land, which she said felt heavy and stagnant. My client had been advised not to let her cat outside until the land was cleared.

"During the sound bath I gave, I called on the support and guidance of angels and other higher realms. The palpably powerful sound bath included intuitively playing all the bowls in the Elemental Set as well as toning with my voice for 30 minutes. Even my cat chimed in, meowing outside my door! I was aware of an overall healing balm for the earth as well as a releasing of trapped souls from the Civil War where the fighting had often been brother against brother, causing deep betrayals and broken hearts.

"When we do a sound bath, we offer it in service, without expecting any particular outcome, although we always appreciate any sign of positive effects. My client noticed a stunningly beautiful sunset afterward, and all four family members slept the best they had since moving in a few months earlier. The following day presented unusually warm, sunny weather and, overall, everything felt more peaceful. The animals happily spent time outdoors, and my client had more energy than she had had in many months. We were both pleased with the outcome, and so were the animals!"

THE MIRACLE ON THE BIG ISLAND OF HAWAII

At the end of 2018, shortly before New Year's Eve, I led an exquisite meditation in Waimea on the Big Island of Hawaii. It was another unforgettable sound-bath experience, with a particular set of bowls playing as they never had before. The music was sublime; it was as if the orchestra was in heavenly harmony, playing with joy and filling every corner of the big studio. One of the bowls was an F note, the note of the heart, and it was infused with Andara, a green volcanic stone that carries the energy of lava, burning away everything that no longer serves.

As I was playing, that bowl was singing wildly. Suddenly a brilliant beam of Light with a huge, golden ball right at its center shot through the room. Whoosh! Twice. Both times from right to left. Like a rocket. I grounded myself and continued to play. My eyes were wide open. The guests were in meditation with their eyes closed. I looked to see if the Light had come perhaps from the headlights of a car shining into the room. There was not a car in sight. I knew I had just witnessed a phenomenon. What had I seen with my naked eye? It was an incredible sight times two!

When the meditation was over, a woman in the front row pointed to the green bowl and asked, "What is that bowl? Why was the sound so strong for me?" I explained it was the Andara, which activates a powerful feeling of love that helps heal the heart. Immediately tears rolled down her cheeks. Reaching into her pocket, she pulled out five small Andara stones.

"I cannot believe this synchronicity!" she exclaimed in quiet wonder. "I have never been to a sound bath, never heard of alchemy-infused singing bowls, and yet I knew I was supposed to come this evening. My baby boy died of an immune weakness shortly after his birth," she continued. "A healer gave me these five Andara stones to work with their energy to help heal my heart. I have been swimming in grief for the past few months, and the sound of that one bowl was so powerful for me, it made me weep. I could feel my baby boy: gentle, loving hands of Light on me. And he was here, my angel. My heart is healing!"

Generosity

Everyone was still. The room was electric. We had all experienced the miracle!

People began to ask me, with awe in their voices, to please explain the two light-beam flashes like lightning bolts, and the balls of golden light like the sun, that had flown across the room. Despite their eyes being closed, they too had seen what I had seen. What was this inexplicable occurrence? And how does it happen that a woman in unfathomable grief decides to attend an event she knows nothing about that ends up changing her life? What was being transmitted through the crystalline vibrations that lit up the room and enabled healing?

Although I have experienced miracles of awakening over and over, I have come to understand that certain things are inexplicable.

As shared by Dr. Carbone, Sacred Vibration is an essential tool and valuable component in the integration of ancient healing practices with modern medicine. My son has taken me on a journey I could never have foreseen; he has led me from the traditional path of classical music and singing on some of the most beautiful stages in the world, to much different "stages" using the crystal singing bowls. This speaks to soul contracts and our human experiences. Although we may not be conscious of our soul agreements until life reveals them, it is now clear to me that indeed, there is a grand plan—a connection from the unseen to the seen, and from the visible back to the invisible. We say yes. Because after all, somewhere and sometime, we agreed to walk our path, and with our yes, we heard the call to live authentically and with generosity.

CHAPTER 12

TUNING

You Are the Instrument

> If there can be a definition of spirituality, it is the tuning of the heart... The tuning of the heart means the changing of the vibrations, in order that one may reach a certain pitch which is the natural pitch; then one feels the joy and ecstasy of life, which enables one to give pleasure to others even by one's presence because one is tuned.
>
> **— HAZRAT INAYAT KHAN, *THE COMPLETE WORKS OF PIR-O-MURSHID HAZRAT INAYAT KHAN, THE SUFI TEACHINGS***

As I developed the format for private and group sound-healing sessions and the curriculum of the Sacred Science of Sound Crystal Alchemy Trainings, it became evident to me that the role our vibrational state plays in effective sound healing is vitally important, as I touched on in Chapter 1 at the beginning of our journey. With my extensive career in music, it was clear to me that it was not just the singing bowls at work, but the human instrument too. And sharing that with people, helping them understand themselves as an instrument, connected the dots for many.

Every human being and every voice has a one-of-a-kind timbre; we each have our own vibrational fingerprint, and no two vocal cords produce exactly the same sound. This is so fascinating. We have a great belly laugh when someone gives a spot-on imitation of a celebrity's unique timbre and style of speaking. It's an imitation of something very recognizable. Each of us has a unique frequency that we emit into the world. As we learn to move the energy stuck in buried feelings, negative thought patterns, and physical pains, and as we reclaim the dismissed and abandoned parts of ourselves, we begin to "tune" ourselves. And as we "tune" ourselves, we come to a stronger expression of health and wholeness. It is important to understand that our own inner sense of being in harmony is crucial, as that is part of the potentially healing "mix" resonating through the combination of the quartz and our voice, whether spoken, toned, or sung.

We must understand that *we* are the instrument that amplifies the sounds of the bowls, setting the powerful vibrations in motion. Paraphrasing the concept of tuning that Sufi mystic Hazrat Inayat Kahn taught, true Spirituality is intrinsically tied to our hearts. As we work with our voice, open our hearts, and align with what is virtuous and good, we tune our human instrument. And by taking this perspective, we are integrating and upleveling ourselves to vibrate at one frequency of Truth, which is Love.

How do we learn to tune the heart and resonate at the frequency of love? It cannot be defined as 432 Hz, 440 Hz, or 528 Hz, or anything in between. It is the love that *you* are—your unique vibrational signature and how you express it. Why is this so important in this field of sound medicine? How do we become that finely tuned instrument? That is the journey. As a singer and performing artist, a voice teacher and then a university professor for many years, teaching in both Germany and the United States, the importance of discovering, identifying, and defining our personal alchemy is at the core of my teaching style, and all of my Crystal Alchemy Trainings address these questions. I knew from experience that focusing on one's unique talents changes the need to compete with any other artist and anchors the idea that each of us has a very individualized life and soul purpose, and a distinct vibrational signature that we are here to express. At every level of training I teach, we dive deep into discovering ourselves as a whole-person, human instrument using proven and effective breath techniques, movement, embodiment, and creative, expansive exercises. We also explore the science of sound, music theory, and the history of sound healing. The learning, exploring, and the fun never stops—for me either!

Throughout this journey, we learn how to tune our human instrument and be a natural expression of love. I smile today when I remember one of my students, a well-respected therapist, who said, "Jeralyn, why would anyone need over 120 hours to learn how to tap and swirl a bowl? What can you possibly teach us?" After the training, she clearly got it: "I had no idea how deep this could go! My mind is blown!"

SOARING BEYOND TIME AND SPACE

I am constantly reminded that working with sacred vibrations and healing music is love in action. During my time as a guest practitioner at Kamalaya, the wellness sanctuary in Thailand, I was privileged to give private and group sound-healing sessions, morning meditation practices, and sacred sound concerts. I learned so much about the practical application of sound and that no matter what someone's mother tongue was or culture they came from (as the guests were from all over the world), the way into the body was through the tuning of the Heart. The incredible years working at Kamalaya anchored in me a stability and knowingness of how to work in a very precise way with the crystal singing bowls that I am grateful for. People came to experience health and wellness, and the bowls took them into unexpected places of release and healing. In the private sessions, I worked with clients experiencing burnout, tinnitus, grief, physical pain, illness, depression, sleep disorders, and stress.

A very unusual situation arose as I was asked to create two unique sound-bath experiences for a private group the owner was working with over a few days. To align with her group's schedule, she had asked me to be punctual and end the sound baths at precisely 12 P.M. each day. The sessions were to be one hour long.

The first sound bath was held in an intimate space on the property. I set up all the bowls, gave a short introductory talk, answered a few questions, and began the sound-healing experience. About five minutes into the sound bath, I recognized with concern that there was no clock in the room. My phone was turned off, tucked away, out of reach. I realized this too late. There was no possible way to know what time it was. I could not reach my phone without disturbing the sacred atmosphere that had been created. I took a deep breath. We lose track of time in a sound bath, regardless of whether we are playing or receiving.

I did not think I would know when it was nearing 12 P.M. I did the only thing I knew to do. I called on my big Angel.

"Dylan, I need your support. Please, can you let me know when it is noon?"

"Okay, Mom, I got you," I heard him confirm in his "heavenly" way. He had my back.

As the hour moved forward, I was completely immersed in the profundity of the sound bath, lost in the endless waves of the Cosmos. I heard a quiet voice whisper, "Mom, it's time." I gently guided the sound bath to a close, wove in the glistening sounds of the chimes, and brought everyone back. As participants were slowly stretching and opening their eyes, I then reached for my phone. It was exactly 12 noon.

I accepted that my life was transforming from sadness and yearning to a level of communication and support that I had not imagined was possible. And I was trusting it all, releasing the control of my mind. People were healing, and so was I. None of it made logical sense anyway. Or did it?

One of my clients in Thailand was a strapping young father who worked at a high-level job at IBM in South Africa. He had never experienced a meditation with crystal singing bowls before. After a 45-minute meditative sound-bath immersion, he stood up and, fully baffled, exclaimed, "What kind of *voodoo* did you do?" He was astonished and tried to express to me what had happened inside his experience. He said he had never been able to meditate, never been able to quiet his mind, and throughout the crystal bowl meditation he was clear of mind, awake, and present... but he had had no thoughts! He was astounded that he did not go to sleep and had been fully present. Following the sound bath, he asked how he could implement this experience in his daily life. He received a few sound-healing recordings and instruction in meditation and breathing, and his life transformed—as did the lives of his wife and two children.

I recently gave a talk and demonstration to a large group on the power of sound healing using the crystalline instruments and my recently published *Crystal Sound Healing Oracle* deck with accompanying audio recordings accessed through QR codes, and one of the guests had never heard the singing bowls. She was amazed at how easily she went into a deep state of meditation, which was new for her. She reported that the following day, she felt stressed and listened to a 12-minute video on the Crystal Cadence YouTube channel. She was amazed when her stress completely disappeared. Another man said he could not believe the sound immersion was 25 minutes. Time had stopped. He was so emotionally awakened that he cried and cried. He asked me to please explain what had happened to him. They were both being tuned, and their hearts opened.

Sound Rx

My colleagues Anders Holte and Cacina Meadu are master musicians of interdimensional sound who never intended to create music for healing, yet this is what their music seems to facilitate. They have worked closely with acoustic-physics scientist John Stuart Reid, creator of the CymaScope, and discovered through a series of scientific experiments conducted at John's laboratory in England that the red blood cell count of viable, living blood cells exposed to their live music increased exponentially. The experiments were completed while they performed 20 minutes of their beautiful composition "Dream of the Blue Whale." Anders is co-conductor of the renowned Lemurian Choir. Their soundtrack *Lemurian Home Coming* is loved around the world.

> Anders shared an experience he had after a sound-healing concert the pair had performed in Germany, which was similar to the one I had after the sound bath at Kamalaya. A manager at the Siemens corporation had come to their event as a tag-along with his wife, and he had been profoundly touched, even moved to tears. He approached Anders and Cacina after the concert and asked them to please explain to him what he had just experienced. He said, "I don't cry. I'm an engineer; I work for Siemens." His mind had quieted its dialogue through their sublime music, and he felt himself as he never had before.

THE TUNINGS OF MUSIC: 432 HZ, 440 HZ, AND 528 HZ

As we learn to tune our human instrument, we must not shy away from the complex—and often controversial—subject of the tuning of a crystal bowl. It's an important element to consider, because a bowl's tuning and its note both determine the specific sound.

Most of the instruments in an orchestra can be tuned to any number of different vibrational frequencies, but harmony, clarity, and the audience's listening experience require that they are all tuned to the same one. The term *hertz* (Hz) refers to cycles per second, and it is used to measure frequency and pitch. In 1939 the International Organization for Standardization officially recognized A4=440 Hz as the standard concert pitch. This means that the A note above middle C should have a frequency of 440 Hz (or vibrate at 440 cycles per second) and that all instruments in an orchestra should be tuned to that frequency. This not only made for better listening experiences, but it also made practical and economic sense. With tuning specifications in place, instruments could be more easily manufactured and sold throughout the world. Pianos made in different countries, for example, would all be tuned to A=440 Hz, and an instrumental soloist performing with different orchestras would consistently expect an A4 note to equal 440 Hz. This is the tuning of almost all music we listen to today, regardless of style.

Tuning

Crystal bowls come with a fixed tuning that cannot be modified. The individual's choice of tuning is a matter of need and preference:

- 432 Hz (representing the A note)
- 440 Hz (representing the A note)
- 444 Hz (representing the A note) creates 528 Hz (representing the higher C note in octave 5), which is a brighter-sounding tuning.

These tuning frequencies only represent the *tuning* of the A or C notes. They do *not* represent the actual Hz of each bowl. The Hz of each individual bowl will always be a different number. You can easily download a tuning app and measure the exact Hz of your singing bowl, if that is information you would like to know.

For example, even though one might expect a C note and a G note to play nicely together, if the C is tuned to 432 Hz and the G is tuned to 528 Hz, they may not sound well together. In music, we would say they are not in tune.

What tuning are you drawn to? Each tuning offers something different, and although there are many opinions about this, in the world of crystal singing bowls, one tuning is not better than another. All are good: there is no one-size-fits-all solution in sound healing. There may be times when you prefer 440 Hz tuning and times when 432 Hz or 528 Hz tuning is more appealing. I invite you to explore the various tunings. Each is unique.

*Go to the QR code at the back of the book
to listen to these three tunings.*

> ## Finding Consonance: Not All Notes Play Well Together
>
> As with all sound medicine, there is no universal prescription for healing; however, a knowledge of musical structure is important. In my experience, a grating or dissonant sound should not be played for a long time or tolerated. If a sound or combination of sounds is not comfortable, it is not the vibration for you. There are moments, of course, in music when a particular harmonic will sit in dissonance before it resolves, and when played with the pure intention of healing embedded in the overall direction where the music is going, it can be effective.
>
> During a bowl consultation, many clients ask for a heart-chakra note—an F—and a crown-chakra note, which is a B note. I could not relax in a sound bath that was built on these two notes. Yet of course it makes sense to combine the heart with the high brain centers. However, the combination of an F note and a B note creates a tritone—or "devil's tone"—in music, which can be very unsettling. It does not create a container of safety or coherence. These two notes alone would not be a good partnership for sound healing. If someone is purchasing two bowls, I would instead combine an F note with an A♯ or a B note with an F♯. This is where knowledge of music theory is helpful. These combinations make the sacred intervals of the perfect fourth and the fifth. I have had clients who ask me to explain why their bowls don't sound well together, and it is sometimes because they have inadvertently bought bowls that were tuned differently, or the combination of notes is not ideal. I have consulted with clients who bought bowls based on their colors or design and were disappointed to hear that when played together, their notes did not make a harmonic sound conducive to relaxation.

In playing the bowls for thousands of people around the globe, I have learned that when you make music with their multilayered vibrations and frequencies, many people experience all their senses lighting up. They may experience a strong connection to God. They see geometrical shapes and colors. They smell roses and other beautiful scents. They taste a bitterness or a sweetness that they either release or savor. And often people have told me they feel physical touches from the unseen world, such as a light tap on the shoulder or hand, or a release of physical pain. The pure quartz with the infused alchemies activates the human system in a way no other instrument does. People feel as though the crystal bowls are turning on an electric current, and an energetic spark lights up in them that they have not experienced before.

BRAIN WAVE STATES

Just as the energy of the person playing the crystal bowls impacts the way the music makes you feel, the individual notes and the alchemies of the bowls also play a role. Personally, I love mixing certain tunings and creating a very special-sounding beat frequency, or what I call "shimmering sound." A beat frequency is not to be confused with a binaural beat, a term you may have heard. A true binaural beat, as Dr. Levitin explained to me, is the unique third tone heard by the brain only when one tone is presented exclusively to one ear and the other tone to the other ear through headphones. Although evidence for binaural beats creating any kind of effect is lacking, the hypothesis is that the beats somehow join the two hemispheres of the brain since different information goes into each hemisphere. In my experience, the beat frequencies or shimmering sounds created with the crystal alchemy bowls have many important benefits. They seem to be exceptionally effective in helping people drop into states of calm, focus, and renewal; deep relaxation; and clarity. We are excited to explore this in a research environment with the newly created Minerva Lab.

When I want to include this unique sound effect, my preference is to use the same note but in different tunings. This might be two different D notes. For example: The D note *tuned* to 432 Hz is vibrating at approximately 288 Hz. The D *tuned* to 440 Hz will be vibrating at 294 Hz. This creates a difference of approximately 6 Hz (please remember that the first Hz numbers represent the musical tuning, while the second number represents the actual Hz of the D notes). This can *theoretically* entrain the brain to reach a coherent brain-wave state, easing stress and potentially restoring us to homeostasis. Remember my client from South Africa, who asked me "What kind of voodoo did you do?" This *is* the "voodoo"! The beat frequencies made with crystal bowls create a unique sound experience because they are produced by acoustic instruments rather than electronic sounds.

As a general reference, I provide this chart with the most commonly acknowledged brain-wave states and their characteristics. However, this subject is still being researched.[1]

STATE	Hz	STATE TYPE	STATE CHARACTERISTICS
Gamma	+ 30 Hz	High-volume/simultaneous information processing (Only brain wave present in all parts of brain)	High coherence Relaxed, intense focus High-complexity problem-solving Spiritual connectedness Bliss states
Delta	5-4 Hz	Deep, dreamless sleep Deep relaxation Decreased awareness	Deep physical relaxation Renewal Access to the unconscious mind
Theta	4-7 Hz	Slow brain activity (between wakefulness and sleep) Meditative state Inner/spiritual awareness Inner focus, creativity, intuition	Promotion of healing Memory and learning Daydreaming Mind-body integration Feelings of bliss and oneness Access to the subconscious mind
Alpha	7-12 Hz	Relaxed alertness and attentiveness Mental resourcefulness Conscious-subconscious bridge	Enhanced, conscious relaxation Meditation Accelerated learning Creative problem-solving Mental clarity
Beta	12-30 Hz	Fast brain activity Analytical mind Reality judgment (Predominant wave when eyes are open)	**Low Beta** 12-15 Hz: Relaxed focus **Mid Beta** 15-18 Hz: Thinking, awareness of self and surroundings. Task performance **High Beta** +18 Hz: Agitation. Fight-or-flight response

> ### Sound Rx
>
> One of my students, Anne, is an experienced Alexander Technique practitioner. Alexander Technique is a popular alternative therapy based on improving posture to improve health. It was created in the 1890s by F. Matthias Alexander, an actor who lost his voice. (I've studied the Alexander Technique since I was 15, as it is an important body-alignment practice for singers to learn.)
>
> Anne shared with me that one of her elderly clients, during her first sound healing, went into a state of deep relaxation. She saw colors and geometric shapes, which astounded her. She asked Anne if crystalline sound healing was a new religion? She could not define what she felt but described with lucidity the amazing depth and clarity of her experience with great astonishment!
>
> She was most likely in the theta brain-wave state. We look forward to researching and gathering scientific evidence to support this.

SOUND-BATH STRUCTURE AND THE BEAT FREQUENCY

I have found it can be very effective to use gentle, shimmering beat frequencies in combination with an anchoring tone underneath that creates a sacred interval. For example, beat frequencies with two D notes combine beautifully with a low G note. Some find this incredibly soothing and regenerating. I have found that many men love this undulating beat frequency, and I have seen it specifically inspire new projects, bring deep relaxation, and shift depression; this technique seems to bring focus and integrate calmness into the nervous system.

In 2021 I presented these shimmering sounds at Vix Camps, an online experience with Grammy Award–winning bass player Victor Wooten and his students, friends, and colleagues. I played three F notes at the three different tunings:

- 440 Hz (a four-alchemy bowl mix of Iron, Gold, Platinum, and Dead Sea Salt)
- 432 Hz (Dylan's Saint Germain bowl)
- 528 Hz (Lemurian Seed and Shungite bowl)

Together they created an intense and gorgeous beat frequency. It was great fun to share this with fellow musicians. They wanted to say, "That F note is too flat. *That* is too sharp!" or, "That one is in tune," basing their observations on music today tuned at 440 Hz. They loved the kind of "bending" sound the three F notes made when played together, which surprised them. I had them close their eyes, breathe deeply, and experience the sound itself without their musician's ear and mind analyzing it. They were amazed by how they experienced feelings of relaxation and focus simultaneously. I intentionally play these sounds in my concerts and presentations, as they are surprisingly beautiful and effective in bringing the audience to a deeply relaxed brain-wave state. And in that state, there is emotional release, recalibrating, and transformation.

Forever Love Accompanied by Singing Bowls

For several years, I shared my crystal alchemy sound-healing concerts in Science and Spirituality events with Gregg Braden and his guests: Dr. Bruce Lipton, Dr. Joe Dispenza, Kryon (the Spirit Essence channeled by author Lee Carrol), Anita Moorjani, Dr. Todd Ovokaitys, and Lynne McTaggart. These concerts brought inspiration, joy, and many memorable experiences. At one of the events held in New Mexico, I arrived at the hotel where I was to perform my *Forever Love* concert with my crystalline orchestra. A very nice young man helped me with my suitcase and all the crystal singing bowls I had brought for the concert.

As I pulled some money out of my pocket to thank him, he said, "My name is Dylan, and it was my pleasure to help you."

My jaw dropped: of all the possible names! I smiled to myself as I looked upward in gratitude. My son was making sure I knew I was not alone.

The following evening, another young man helped me bring all the bowls to the ballroom to set up the stage for my concert. As I went to thank him, I saw his name tag. It said "Dylan." I laughed and told him the young man who had helped me when I arrived also was named Dylan.

> He smiled. "It's so unusual," he said, "but we have four Dylans working here! Three are spelled D-Y-L-A-N, and one is spelled D-I-L-L-O-N."
>
> I marveled. "My son is named Dylan!" I exclaimed.
>
> I got the message loud and clear: "Mom, I'm always with you. And four of me, like a quartet! We are tuned and in Divine Harmony! Let's have fun! Green Light, Mom!"
>
> That was more than a coincidence. The bridge connecting Heaven and Earth never ceases to amaze me. We are energy, made of frequency and compressed light, eternal beings. As we tune our hearts, we traverse that bridge on the wings of sound, in joy, and grounded in Love.

MUSIC TUNES THE HEART

I have learned to navigate grief and the gamut of emotional triggers with love. It's what kept me here and keeps me here. Forgiveness. Compassion. Kindness. Understanding. Dropping deep inside and finding a place of complete stillness. Knowing that pure, healing tones are made from that stillness, and that inherent in light is always darkness. Inherent in darkness is always light. When we embrace that darkness, it reveals the potent power and possibility of the light. Dylan's communication is clear, and he is busy!

"Music and sound vibration is our medium, Mom, bringing healing."

"Mom!" (His voice is still loud like his mom's.)

"Mom, tell them Love wins."

"Mom, tell them!"

"In the face of the unfathomable, in the face of chaos and confusion, in the face of whatever life delivers to us, both young and old—loss, grief, addiction, illness, injury, pain, abuse, regret, depression, disappointment, death. . . whatever it is that puts us right in the face of the bitter cold, show them how to find the warmth of a thousand suns in their own belly. Mom, show them how to turn their faces to the Light!"

Long exhale. . . understood, son. Turning to the Light. And Tuning the Heart to the frequency of Love. It is a choice. I'm walking that path of Light with my beloved son—not as a parent would choose, but then again, in some cosmic whisper of reality, I had agreed.

CHAPTER 13

AWARENESS

Embodiment and Wholeness

> Blissful, oh beautiful is your Divine World! Lift me
> to your world of a million twinkling stars!
>
> **— AMMA (MATA AMRITANANDAMAYI DEVI)**

I looked directly into the Hugging Saint's kind brown eyes, and the tears welled up and spilled over. I was in a small fishing village called Parayakadavu on the coast of southern India, where I was attending Darshan at Amritapuri, the ashram of Amma. This beautiful humanitarian and Hindu spiritual leader had comforted more than 34 million people across the globe. That day in August 2016, I became one of them. I had traveled almost 10,000 miles from Los Angeles to India—first to Delhi, and then south to Parayakadavu—to meet Amma. Deep in grief, I had been visiting with one of Dylan's closest friends and his family, and his mother and I made the pilgrimage. It was unlike anything I had ever experienced or anywhere I had ever been. People from all around the world—with different languages and cultures—had gathered in union and in the spirit of Love. Here was a living example of the ancient Indian ideal that the whole world is one family. We stayed two nights at the ashram, a place emanating humbleness and healing. The first evening Amma chanted, and we received her meditation and her powerful mantra music. A gentle peace permeated the room and floated lovingly through the large crowd.

I held Dylan's small picture, the one that had the quote from his college application written on the back, and that Dr. Sue had taken to Huayna Picchu, the high peak at Machu Picchu. I planned to share it with Amma and ask her to bless it. When I arrived at the front of the line, our eyes met and Amma hugged me; her assistant translated the story of Dylan's passing, my grief, and my journey to India. And then I saw something happen I had never seen the likes of before that moment. Amma set Dylan's picture on the small table next to her chair and placed her index finger right at his brow point, his third eye. A ray of light emanated from her finger to the photo directly on the sixth chakra point. I saw it with my own eyes. A ray of light came through her body, radiated out of her fingertip, and extended into Dylan's photo—directly to his ajna chakra, the place of intuition and insight, inspiration, and vision. I felt the electricity of the light beam come straight through his photo and simultaneously, I received it inside of me. I cried in wonder. Amma looked at me with great love and motioned for me to take a seat beside her on the right, where I sat for the rest of the evening.

I remained in this sacred space in complete awe until the event was over and we went quietly back to our rooms. I slept deeply. I connected with my son. The grief shifted that night. Early the next morning, I awoke with a new sense of trust. Sharing with my friend Bhavna, I found that she too had seen and marveled at the light beam that had shone from Amma's fingertip. There was something beyond my comprehension happening here that was unfolding in front of my very eyes. And yet, somehow, I understood it. This was healing in the present tense. I had an unbelievable journey with Spirit when I was at Amma's ashram and experienced what Dylan kept confirming through his communication.

"We are made of energy, sound vibration, and Light. It is where we come from and where we will return."

These words have become my mantra, alive and well in every cell of me.

After this visit to India, I was quickening, embracing life in a new way, and crystalline music was the container for it all. My mother stood solidly behind me; she was my biggest advocate. One day, half teasing and half questioning, I said to her, "Mom, what has become of me? Here's this classical musician, singer, professor of voice, and now I'm talking to my son on the other side and playing these crystal bowls. Really? Mom! Have I lost my marbles? Who am I now?"

She said, "Dear, no, please do not question yourself. So much good is happening. You are healing, and you are helping so many people."

Bless my mom. She has given me life; nurtured, held, and supported me; and seen me through the life and death of my precious child and into my own rebirth.

I treasure the fact that my disciplined academic and artistic backgrounds are now merged with the esoteric, with ancient wisdom and the light of other realms, and through my journey with Dylan, the healing power of music and the singing bowls, I continue integrating and embodying the wonder of it all. I am living the ray of light, the connection of Heaven to Earth I saw come through Amma's fingertip. And I know this is available to everyone.

It all makes tremendous sense to me now, and the miracles of communication without language confirm this realm of the sublime, the wordless, that I knew so well from my career in music, but not at this pointed level of spiritual awareness. And yet through bio-energetics and sound vibration, I was learning to embrace what I somehow sensed my whole life: There was more and *I was actually living it*, this connection to other dimensions, and *it was real*. I saw evidence of this at every turn, and recognized that by being willing to feel the unfeelable, I was elevating my frequency. Music, especially crystalline sound, helps healing on all levels of being: on the physical, the emotional, the spiritual, and on the soul level. There is an energy the sound carries that transforms. How do you describe something that is essentially invisible? It is the language of the heart, the vibration of pure love: no science can fully explain it with all its mysteries. The high-frequency crystalline sounds help us transcend the thinking mind connecting us to the Soul, eternal and immutable.

As I walked the path of transforming my grief and rediscovering joy and purpose, I learned how crucial the principle of embodiment was to the upleveling and sustaining of my new vibrational frequency. If I could not integrate and ground what I was learning and experiencing, I would quickly fall back into hopelessness and depression, and in that state, Dylan could not communicate with me—we would not be a vibrational match. The motivation for integrating and embodying my newfound frequency was to remain in contact with Spirit, with a higher vibration.

The weekly sound-healing immersions and classes I was teaching were rapidly expanding. I continued my travels and personal healing journeys in combination with service and the sharing of sacred sound.

THE IMPORTANCE OF EMBODIMENT

Before I begin a public sound bath, wherever I am in the world, I often ask everyone what the pertinent themes in their lives are. I also ask what their intention is and what they would like to receive. People regularly ask for help with opening their third eye, and many participants want to experience a psychedelic opening through the aural sensations of the sound baths. I teach them first about embodiment.

Through the third-eye experience on Dylan's photo, Amma had given me a potent message that had activated a strong sense of responsibility as I taught people the bowls. An out-of-body experience is indeed a worthy goal, yet I had learned how important it is to be comfortable residing in our physical form. As we create expanded experiences of light and bliss, connecting to other dimensions, we must know how to come back down, return to our humanness, and live at home in our physical body. We need to know how to integrate these high-frequency experiences to truly elevate our personal vibrational frequency. We must honor the value of integration and grounding, for when high-awareness experiences are integrated and anchored in the body, they can be repeated. They become sustainable. Bliss becomes a natural, recallable part of you. With these tools, we can release physical pain and become more emotionally balanced.

How do we ground? We become aware of the flow of energy streaming through us, filling any emptiness we feel. Without judgment we allow the movement of energy within us. A committed wellness practice supports grounding and can include the integration of conscious breathing, the power of intention and presence, and the unimpeded flow of energy up and down the central channel of the body. Movement is an important aspect of grounding. Tai chi, qigong, yoga, breathwork, heel strikes, deep squats, conscious exercise, cross-lateral movement, and dance all support integration and grounding. Also important are meditation, prayer, the attitude of gratitude, singing songs you love, toning, humming, mantra, and taking time to be in nature—feet on the earth and experiencing the sacred connection to all that is. This is oneness. This is bliss. And this is the feeling you want to nurture, repeatedly, to sustain a high vibrational frequency. Beyond these practices, it is imperative that we are vigilant with ourselves about staying aware of any areas of discomfort and being committed to work with the raw energy to release any blockages. We cannot shy away from our pain. The invitation is to practice and discover how you personally ground and embody. Remember, you are an energy being and always in vibration. There is no one way to ground, yet it is the key to health and vitality; and sound vibration is an integral component of rootedness.

BEING THE SOUND: A SINGER'S INSIGHT INTO EMBODIMENT

The HeartMath Institute, a nonprofit organization dedicated to scientifically validated tools to help awaken the heart of humanity, teaches that our brain receives impulse(s) from the heart, the heart from the earth, and the earth from the sun, encouraging us to master a higher vibration than thought. And why? Because thought alone is not the whole picture of who we are. It's like my singing teacher said so many years ago: "Stop listening to the sound of your voice. Be the sound. Get out of your head; nothing vibrates there. You must learn to listen and sing through the feel of your body."

It took me several years to accomplish this, but I remember exactly the moment when it happened. It was a concert performance in the Cathedral of Saint John the Divine in New York. It was pure bliss. I let go of my critical mind and anchored the feeling of the flowing expression of my voice in my body, so I was able to repeat it. I recognized that, like an athlete, I was no longer thinking, listening, or judging myself: I was singing from every single cell of my body, completely grounded. My heart was activated and "tuned." I felt a strong sense of embodied confidence and joy. In that moment I understood what my teacher had been sharing with me: that *physical experience* was a higher vibration than thought. This is embodiment. This is true Mastery of Technique—"feeling the feeling" vibrating inside of you and being able to recall it and repeat it.

A few years later, I stood on the stage of the concert hall where the Munich Philharmonic orchestra plays, performing the jump-in that included the *Tosca* aria I mentioned earlier in this book. At that moment, this concept of expressing from the inside, being embodied, was the key to trusting in the moment that all I needed was available and inside me. I was a vocal athlete, poised and ready to play the game, the music of the heart.

I had another intense experience when I was honored to sing the national anthem in my hometown of Los Angeles at a Lakers game at the Staples Center for a crowd of 18,000 people. There was no rehearsal. The halftime show rehearsal had gone overtime. I was handed the microphone and walked out to the center of the basketball court. As the phrase "and the rockets' red glare / the bombs bursting in air" sailed out of me, the crowd started booing. It took all my inner strength to stay present and finish the anthem, and I remained in my core, singing from a grounded, embodied state. But why the boos? I walked off the court wondering as a very handsome man in a rainbow-colored beanie sitting courtside jumped up and took my hand.

"That was one of the best renditions of the anthem I've ever heard," he exclaimed, "ranking right next to Marvin Gaye!"

It was actor and producer LL Cool J, and I was honored to receive his praise. As I returned to my seat, my family explained that people in the crowd had booed at the words "rockets' red glare" because the Lakers were playing the Houston Rockets! I exhaled. What a relief!

And now, every time I play the bowls, I am clear. I am embodied in the technique I mastered as a singer, and *every* cell in me is playing too.

Sound Rx

Music. . . will help dissolve your perplexities and purify your character and sensibilities, and in time of care and sorrow, will keep a fountain of joy alive in you.
— Dietrich Bonhoeffer, *Letters and Papers from Prison*

The personal expansion the bowls can ignite is available to all of us, no matter who we are or where we live. Energy worker Jessica Neideffer began offering sound healing and self-awareness classes to prisoners through her nonprofit organization, the Adara Collective, in August 2022.

"The men, women, and children I've shared space with inside these correctional facilities have been so open and vulnerable that at times, they bring tears to my eyes," she says. "Not because I feel sorry for them but because I get to witness them remembering who they truly are. The level of vulnerability and consideration is impressive and inspiring. The individuals share from the deepest places within their Self, and they do it without shame more and more with each session. As they begin to transform, they open to their gifts, and the amount of talent and creativity you experience among these individuals is amazing."

Jessica finds that during the sound baths, participants are able to forget their physical space for a while and play in their imagination. They report that this helps them adjust their perspective on what *is* and *is not* under their control.

> "The participants share about the peace and calm that washes over them like they've never felt before," Jessica notes, "and the curiosity and questions that come after are always insightful. People feel seen and heard, sometimes for the first time, which supports them to feel confident in themselves and trust how they're feeling.
>
> "The therapeutic use of sound for healing within our prisons not only supports the rehabilitation of the individuals housed there, but the officers and staff benefit too," Jessica continues, noting that staff have said that they find literature from the classes during bunk searches and that people tend to discuss the information they've learned at Jessica's classes when they return to their dorms. The officers have also told her that the people who attend the sound healings are usually the ones who help to defuse potentially aggressive or violent situations.
>
> "Overall, people are less angry, more peaceful, and less bothered by others or focused on the limitations of their current circumstances," Jessica adds. "In fact, the officers have become so curious about the sessions that they are asking for support in this way as well. . . . There is so much potential for positive change within these spaces through this modality and our supportive intentions!"

THE TRUTH OF WHO WE ARE

There's no doubt that whatever groups and demographic we are playing for, the crystalline music received, transmitted, and amplified through our human quartz-like structure, takes us beyond our thinking, analytical, judgmental and sometimes critical mind by grounding us in the higher, exalted vibrations. Using sound, we awaken, increase our awareness, and land in the truth of who we are. Pure sound and music with intention have the ability to stabilize, center, and expand us, guiding us beyond our beliefs, stories, and genetic inheritance. Using both our own voices and the crystalline instruments, we can physically feel the resonance and vibrations and discover:

- sounds that ground us, bringing a strong sense of stability.

- sounds that center us, bringing a sense of safety and home.
- sounds that accelerate us, helping expand our consciousness and ability to connect easily with the higher realms.

With these three energies present, we become vibrating chords. I liken it to us being an essential part of Nature. The different depths of sound blend the elements of water, earth, and sky, and we become the song we are here to express.

CHAPTER 14

STRUCTURE

The Impact of the Crystal Singing Bowls on the Human System

> There is a light that shines beyond all things on earth, beyond us all, beyond the heavens, beyond the highest, the very highest heavens. This is the light that shines in your heart.
>
> — FROM THE *CHANDOGYA UPANISHAD*

Great music is built on solid structure and enhanced through the mastery of technique and emotional expression. Nuances are communicated through the talents of the artists performing, be it the conductor, the orchestra, the instrumentalist, the band, or the singer. I learned this early in my profession: mastering technical skill is essential, yet equally important is your unique creative expression. I first heard Beethoven's opera, *Fidelio*—which I was to sing often in my career—performed at the Zurich Opera, with two different casts and two different conductors. Under the baton of each conductor, I experienced the brilliance of Beethoven's musical structure. Yet there was a remarkable difference in the subtleties of expression. At certain moments it did not even seem like the same composition. I was truly astounded at how different each performance sounded and felt. This cannot be the same piece of music, I thought. One conductor was passionately into the music; one seemed to be performing routinely.

Why is this story relevant? In the world of sound healing and crystal singing bowls, although it is not necessary to be a trained musician, as in any other profession, knowing your subject and mastering the skills is indispensable. This means having a thorough understanding of certain structures. Here we will explore some basics. Thousands of years of sacred tradition and the theory of music will be our guide.

THE CHAKRA SYSTEM

Chakra in Sanskrit means "wheel" or "disk," and it is commonly used to describe the seven main energy centers located along the spine:

Root, or muladhara chakra
Sacral, or svadhisthana chakra
Solar plexus, or manipura chakra
Heart, or anahata chakra
Throat, or vishuddha chakra
Third-eye, or ajna chakra
Crown, or sahasrara chakra

These energy centers have a resonant affinity with the frequencies of the crystal singing bowls and their music of the higher dimensions. It is important to understand that emotions are energy in motion, and the sounds of the bowls create a safe container for our emotions to be felt and expressed. Their coherent, high-frequency harmonics can allow us to experience awakenings and consciousness expansion as well as a deep, earthly connection. In other words, they help us fly with our feet on the ground. There is something inherent in the sound of the crystalline instruments that connects us to what Beethoven describes as the "electrical soil where our spirits live, think, and invent." The pure tones of the bowls evoke soulful memories in the core structure of our cells and bring us back to balance. Their harmonic essence may restore our sense of belonging and unity.

The seven main chakras represent a level of consciousness and different aspects of life. Among other things, they are associated with planets, colors, organs, elements, musical notes, and vowel sounds or seed mantra. These powerful sounds are traditionally used in yoga and meditation.

In the ancient metaphysical system of chakras, the concept of sound, or nada, is based on the premise that the entire Cosmos consists of vibrations, as we explored in Chapter 5. These vibrations are divided into two musical categories: the internal, silent vibrations, anahata, and the external, audible vibrations, ahata. Throughout this book, we are rediscovering the power of our internal music. This is the music of anahata, the music of the heart, where all of the sounds of each chakra emerge. (See the chart on page 167.)

THE SEVEN TRADITIONAL CHAKRA NOTES

In the Sacred Science of Sound Trainings, we work with some additional chakras that include the North Star, representing our life purpose, and the Earth Star, representing a grounding of our legacy. Together, these chakras function as a connector of Heaven and Earth. The life purpose chakra (or sutara in Sanskrit) is located in the energetic space two feet above the head, and the legacy chakra (or vasundhara in Sanskrit) is located in the energetic space two feet below the "souls" of the feet. Just like the other chakras, they have their own functions and corresponding musical notes.

Chakras are a gateway between energy and matter, body and soul. Sound vibration can tune and activate the invisible nature of the chakras, allowing us to express our unique purpose and potential while supporting us to live fully in the giving and receiving flow of life.

THE NOTES

Let us explore the relationship between notes and chakras. For our purpose here, we will be using the Western notation system: C, D, E, F, G, A, B (or do, re, mi, fa, sol, la, ti, as you might have heard in the famous song "Do-Re-Mi" from the Rodgers and Hammerstein musical *The Sound of Music*. These are the notes of the C-major scale.

As represented in the graphics below, you can see that the Legacy chakra has a lower-octave G note than the throat-chakra G note, and the Life Purpose chakra has a higher-octave C note than the root-chakra C note. It is important to know that these "low" and "high" notes function as part of a whole, just as the two poles of an electric current are equally necessary to create one light. Elevation is not sustainable without rooting.

This illustration shows the notes of nine of our chakras on the piano keyboard:

**CHAKRA TONES
INCLUDING LEGACY AND LIFE PURPOSE**

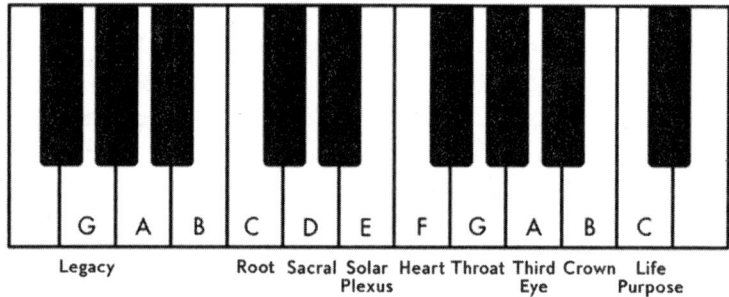

When we work with sound in combination with the nine energy centers represented here, our conscious breath, the principles of embodiment, and our loving presence, we keep the chakra system balanced, helping us stay healthy and whole, and open a pathway from the inner to the outer realms, from the finite to the Infinite. Sound becomes a bridge, a medicine for grounded expansion.

CHAKRAS: THE SEVEN WITHIN THE BODY AND TWO OUTSIDE THE BODY

Chakra	Area of the Body and Main Organ	Note	Color and Affirmation	Element	It Represents:	Bija Mantra Sound (Sanskrit and Western Vowel Sound)
Root or *Muladhara* Chakra	Base of spine, reproductive organs, legs, bones	C	Red I am safe and I am home	Earth	the physical body; the sense of being rooted, of belonging	– Lam – uh
Sacral or *Svadhisthana* Chakra	Below the belly button/ lower abdomen organs, adrenals and kidneys	D	Orange I am the source of creation. I am wisdom rising	Water	the emotional body, creativity, and passion	– Vam – oo
Solar Plexus or *Manipura* Chakra	Behind the navel/ pancreas, liver, digestive system	E	Yellow I am confident I am joyous	Fire	the mental body, personal power, and self-esteem	– Ram – oh
Heart or *Anahata* Chakra	Thoracic cavity/ heart, lungs, thymus, hands, circulatory and respiratory systems	F	Green I am love and compassion	Air	connection, unity, compassion, and unconditional love	– Yam – ah
Throat or *Vishuddha* Chakra	At the throat/ thyroid, parathyroid, neck, shoulders, vocal cords	G	Blue I am Truth I am Grace	Ether	communication, both speaking and receiving, and manifestation, space	– Ham – eye
Third Eye or *Ajna* Chakra	Between the eyebrows, midbrain/pituitary glands, medulla, eyes and brain	A	Indigo I am intuitive I am perceptive	Consciousness	vision, intuition, perceiving	– Om – aye (as in bay)
Crown or *Sahasrara* Chakra	Crown of the head/pineal gland, cerebral plexus, upper head, whole body	B	Violet/white I am connected to all that is. I am bliss	Light	unity, bliss, and collaboration with a higher realm	– Silence – ee
Life Purpose or *Sutara* Chakra	Two feet above the head, unity consciousness, Heaven/Earth connection	High C	Golden white I am the higher perspective I see the bigger picture	Light	The bigger picture; the bird's-eye view of life, your unique purpose or North Star	– A high *whooo*
Legacy or *Vasundhara* Chakra	Two feet below the body/earth, connection, legs, feet, bones	Low-octave G	Terra cotta I am grounded in my unique purpose I live my spiritual inheritance	Earth	Embodiment of your Life Purpose; your Legacy or Earth Star	– Deep tone of *om*

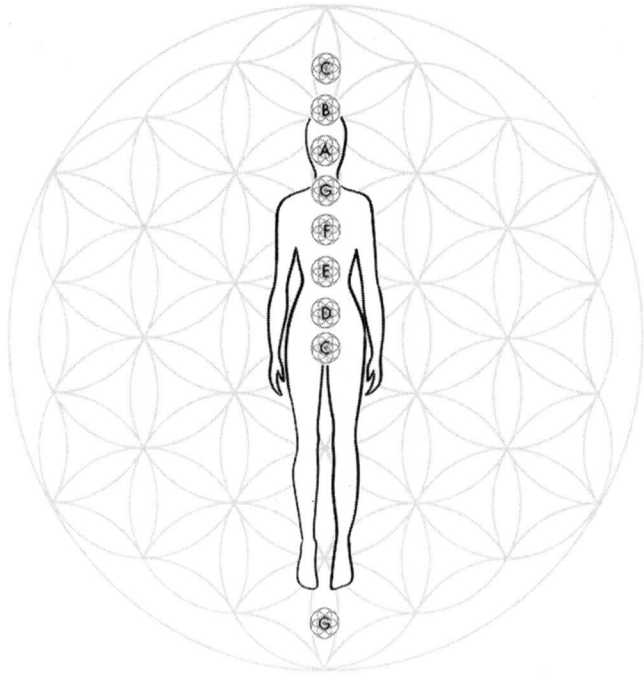

The seven chakra notes on the body plus the additional notes associated with two of the chakras outside the body.

THE POWER OF ANAHATA

As Swami Satyananda Saraswati beautifully says in his classic book, *Asana Pranayama Mudra Bandha*: The word anahata literally means "unstruck." All sound in the manifested universe is produced by the striking together of two objects, which sets up vibrations or sound waves. However, the primordial sound, which issues from beyond this material world, is the source of all sound and is known as *anahad nada*, the unstruck sound. The heart center is where this sound manifests. It may be perceived as an internal, unborn, and undying vibration, the pulse of the universe.[1]

These words are alive with a joyful resonance: this is the essence of sacred sound, the sound that we don't even hear. I will always remember the tall, slender Indian man who had just finished his chemotherapy. He was weakened and very frail. I had put together a spectacular sound meditation with a set of bowls in various octaves of the five black notes on the piano. This creates a pentatonic scale. It was deeply grounding, centering, and expansive. When the sound bath ended, this man could not contain himself! Despite his physically weakened condition, he jumped up from his reclining position, filled with strength and enthusiasm, and shared that he had been completely immersed in memories of his childhood, which he had spent growing up in Mumbai. During the sound experience, he had felt himself on the roof of his family home with his mother and his six siblings.

He described feelings of warmth related to his close-knit family and the palpable love of his mother. It had felt so real that he had shouted! He was filled with pure joy and took this memory to his heart. He stayed after the group session, and he shared with me more details of his experience. We worked together until he anchored and integrated his unexpected feelings of elation. What remained was an incredible sense of freedom and lightness of being.

I watched as he radiated pure joy, this beautiful man carrying all the worries and discomfort of his disease. Five years later (no pentatonic pun intended), I discovered he was still doing remarkably well and living a healthy life. His experience confirms there is indeed a place within us that remembers and longs for this deep sense of natural order and perfect harmony, and this place is the heart. These primordial sounds are embedded in our cellular memory, and the vibrations of the bowls took him to that undying connection beyond time and space.

THE PENTATONIC SCALE

The pentatonic scale owes its name to the fact that it uses five-note combinations instead of seven (such as we just explored with the chakra system). There are different pentatonic scales that use variations of note combinations. Working with these two systems of harmonic combinations in sound healing gives us endless possibilities. The relevance of the pentatonic scale goes even beyond these possibilities. In 2008, when I was living in

Germany, scientists announced the discovery of flutes made of bird bones. This was an unexpected find, as instruments tended to be made from stone or bone material from other animals. Perhaps by using the bones of the winged ones, the ancient artists imagined the birds were imparting their melodious sounds to their precious handmade wind instruments. The instrument relics found in Germany were assessed to have been made 40,000 to 60,000 years ago and interestingly, many of the bird bone flutes were tuned to the pentatonic scale!

What does that mean to us? Historically, it is said that the tones of the pentatonic scale came from the Pythagorean theory of the "music of the spheres" based on the idea that there is music and harmony in the universe. It is reported that during that time, the Pythagoreans wore pieces of five-pointed jewelry with the word *health* inscribed on them in Greek. But we know now the pentatonic scale was used many thousands of years *before* Pythagoras. One of the first records of the use of the pentatonic scale in medicine dates back to ancient China. There it was believed that music had the power to harmonize a person in ways that medicine could not. Curiously enough, one of music's earliest purposes was for healing, to the point that up until today the Chinese word, or character, for medicine or cure (藥) comes from the character for music(藥) and can be traced back to 2650 B.C. Another interesting fact is that music and happiness (藥) are also represented by the same character, demonstrating that this ancient culture had a deep knowledge of the role sound and emotions played in our health.

In Chinese medicine, the pentatonic scale is directly related to the harmonization of the body's five elements through music (Earth, Metal, Water, Fire, Wood) to promote happiness and therefore healing.

A pentatonic scale can be played in any key, starting with any note, and using it with music medicine, there are many possibilities. Well-known contemporary songs such as "The Shape of You" by Ed Sheeran, "My Girl" by the Temptations, and the spiritual song "Amazing Grace" are all based on the pentatonic scale.

A pentatonic scale has a very particular sound. The major pentatonic scale, which is what I use most often, is made of the following musical intervals (see the Glossary): whole step; minor third; whole step; and whole step.

Both the 432 Hz and the 528 Hz downloadable meditations included in this book use a pentatonic scale.

FAVORITE BOWLS

I am always hard-pressed to choose my favorite bowls, because I've never met a bowl I did not like—but I do like some better than others. If I had to choose, I would choose a set built with the pentatonic scale. Why? Because there is a fascinating significance to the pentatonic scale and the number five upon which it's formed. The sacred geometrical shape of the pentagon is Nature's most powerful expression of the golden ratio—1:1.618—known as *phi*—which British acoustic-physicist John Stuart Reid confirms is prevalent throughout all of life, including in the cross section of the DNA molecule. Leonardo da Vinci's Vitruvian Man (see page 172) also contains this ratio, as do certain symphonies of Mozart or works of Debussy and Bartók.

The thousand-year-old practice of yoga applied the principles of sound to the chakra system and understood that each sound had a corresponding geometry called a *mandala*. These geometries represent the creation principles of Nature, which would become the yoga asanas or body postures we use today to bring health and harmony. Throughout history the number five has symbolized the human being. We have five fingers on each hand and five toes on each foot. A good example of this is *utthita tadasana* or "five-pointed star," which among other things lengthens, opens, and infuses vitality into the body.

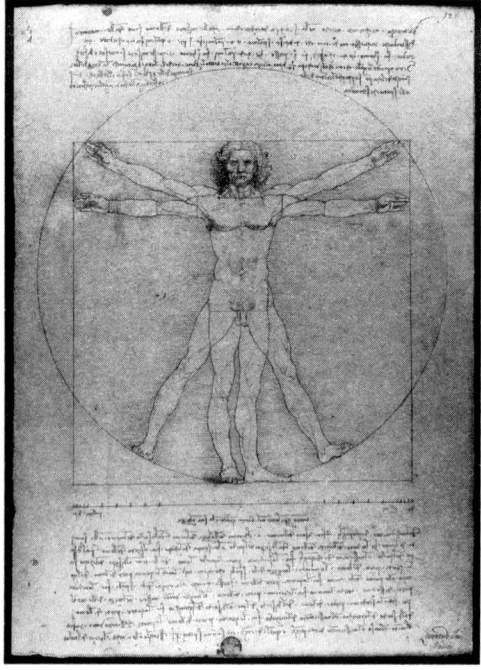

The golden ratio, found in the sacred geometrical shape of the pentagon, is also expressed in the Vitruvian Man by Leonardo da Vinci. Was he a fan of yoga, doing *utthita tadasana*?

The number five connects us to a primordial wisdom that ancient cultures knew well. A good example is the bird-bone flutes mentioned earlier in this chapter. The combination of the five notes of the pentatonic scale seems to be part of our cellular memory and genetic makeup. There is a fascinating video on YouTube with the great vocalist and improvisor Bobby McFerrin (who I had the good fortune to study with) as he demonstrates the power of the pentatonic scale from the World Science Festival in 2009. His interaction with the audience stunningly illustrates this. I encourage you to watch it.

Anton Dvorak's beloved aria "The Song to the Moon," sung in the Czech language, is composed in a musical key that uses all five black notes on the piano. When I presented in Australia at the first Crystal Bowl Symposium in 2017, I sang this aria, infusing crystal singing bowls into the stringed orchestra accompaniment. One does not need to understand the words, and wherever in the world I have sung this it has always created a transformative music-as-medicine experience. While I was in Australia, I had the privilege to meet with some Aboriginal musicians who shared with me that the pentatonic scale was also a familiar pattern of notes in their culture, and they sang the five-note scale to the accompaniment of their native instruments.

The transformational power of music weaves its magic through my life. I have been blessed to play the bowls with many talented musicians.

Making Music with Jahnavi Harrison

Adding alchemy singing bowls to the music of sacred mantra singer Jahnavi Harrison and creating with her the album *Balm* was an experience I treasure. We experimented with different musical keys and with the alchemy bowls she felt good singing with. We added secondary instruments, such as her violin and harmonium, and a HAPI steel-tongue drum in D minor. Jahnavi grew up in England at the Bhaktivedanta Manor in a musical family, and she trained in both Indian and Western styles of music. Her singing is luminous, so it was not a surprise to discover that her favorite key was A, the key of the third-eye chakra, and her second favorite was B, the key of the crown chakra. We added various combinations of notes for the key of A that included a low A note and a high A note an octave above it, plus the notes of D, E, F♯, G♯, and C♯. This is a perfect example of choosing a key matching to the singer and combining both black and white notes.

Listening to Jahnavi's devotional music, you understand how those two keys are perfect. We combined the etheric yet grounded and centering sounds of the bowls in two musical keys that activated inner wisdom and vision, light, and bliss—all that she and her music so radiantly embody. These grounded notes of the upper chakras (crown and third eye) held a magnificent container for her to express her unique vibrational signature. And this is what it is all about: allowing your personal essence to flow freely. The alchemies added yet another color to the orchestra. Combining Mount Shasta Serpentine, Sedona Red Rock, Amethyst, Black Tourmaline, and Platinum brought additional energies of groundedness, balance, healthy boundaries, creativity, and so much more.

The bowls were 440 Hz, as we needed to tune with her instruments—the harmonium and the violin. Setting the intention of creating high-vibrational, devotional, sacred healing music, we opened our hearts and improvised. How fulfilling to be that vibration higher than thought! We had no plan for where our music-making would lead us. I had the technical structure in place, she expressed her mastery, and we met in the unknown, creating from that place of supreme trust. I invite you to listen to the special music of the *Balm* album and to allow it to uplift you to the sublime.

Music Theory for Crystal Bowls: Sharp or Flat?

In the theory of music, the black notes can be named sharps or flats, depending on which musical key you are playing in. One interesting thing to understand about the music theory of the crystal bowl world, however, is that a black note is never named a flat note. Interesting. I imagine this came to be because sharps raise a note and flats lower a note, so naming the black notes "sharps" instead of "flats" gives the musical indication that we are raising our vibrational frequency through sound healing.

An aside: due to the prominence of the # sign, also called a hash mark and which is used to signal a hashtag, sometimes people on social media will ask, "What note is C hashtag?"

"C sharp," I say and smile.

The black keys on the piano have become popular, even trendy—a hashtag kind of thing!

Playing with Victor Wooten

Playing bowls with the legendary five-time Grammy Award–winning bass player Victor Wooten was a very different experience. I asked him what his favorite key was. Victor is a consummate musician, a master. He said, "Any key." We then chose the key of E♭ (the key of D♯ in bowl language). The notes support calmness, helping relieve feelings of fear, worry, or anxiety. With an unexpected mix of Victor's brilliant musical gestures on his Yin Yang bass by Fodera, integrated into a tapestry of crystal singing bowls, this piece took us on a journey of inner peace, poise, and creative flow, working subtly with the adrenal glands, helping to ease fear and the rush of adrenaline associated with fight-or-flight reactions. (This meditation can be found on Source, the app of the Sacred Science of Sound, which is available for iPhone and Android.)

Music with Kevin James

I have made crystal singing bowl recordings in different musical keys for Kevin James, the Australian chant leader. He uses these accompaniments to close his HeartSongs circles while he plays his magical handmade flutes. I created sets in the keys of D (for the sacral chakra), E (for the solar plexus chakra), and F♯ (for the heart chakra and immune system). The combination of bowls and flutes makes for an exquisite ambience of deep relaxation and regeneration. The participants are able to fully integrate their chanting experience, and they leave Kevin's sacred circle filled with pure embodied bliss.

BE OPEN TO WHERE THE SOUND TAKES YOU

A lot of people ask, "How will a sound bath affect me?"

Sound healing works with the subtle energy systems, and everyone responds to a sound bath differently. Be open to where the sound takes you, and always notice what you feel and where in your body you are activated. That is where you are being guided to bring your loving presence, breathe into any tensions, feel, and release.

Please remember that although each chakra resonates with a specific note, all body systems are intrinsically connected and continuously influencing one other. The tones of the alchemy singing bowls do not always translate one to one, meaning you may play or hear a D note infused with Salt alchemy or Lepidolite alchemy, and you may feel it in your throat, while someone else experiences a general relaxation or an emotional release or an epiphany, and someone else actually feels it in the sacral chakra, the very area with which the D note is associated. The alchemy of a crystal bowl, the tuning of the bowl, and the size of the bowl, in addition to its actual note and, of course, the player, are all factors that contribute to the grandness and complexity of these sonic tools. Healing with sound is multidimensional, and we must work with sound in a manner that may sometimes be beyond the grasp of our logical, thinking mind. The information contained here is a guideline. With the increase of interest in sound healing in the past few years, I am a firm believer that a solid understanding of structure, paired with integrity and presence, are uncompromisable factors that will determine the difference in the outcome of the experience—just like the one I had in Zurich with the two different conductors. Sound operates in the realm of the quantum and in collaborative combination with the personal alchemy of the player and their understanding of the theory, history, and the science of sound. This is the world we are exploring, gathering anecdotal information in, and researching. It is vast and filled with possibilities. The alchemy sound teaches us to listen, feel, see, sense, express, and to trust.

As Sound Therapists and Musicians, I feel it is our responsibility to play from our authenticity and never by rote. An interesting anecdote that further speaks to this comes from a private conversation with my friend Dr. Sole Carbone, who was privileged to converse with Dr. Masaru Emoto's Hado Life Water Laboratory director and co-founder Rasmus Gaupp-Berghausen in 2020. According to Dr. Carbone, Gaupp-Berghausen shared with her that when experimenting with the effects that emotions have in the structure of water, the scientists were fascinated to learn that unless the person performing the experiment was able to actually *feel* the specific emotion—for example, compassion—the water crystal would *not* form. In other words, the water was being affected not only by the word that represented the emotion, but by the cellular signal that the heart of the technician was imprinting and translating into the physical molecular structure of the water. Our emotions are our personal vibration, and they affect matter. Expanding on the concepts of Dr. Emoto's work covered in Chapter 6, this intimate story that Dr. Carbone shared with me further attests to the power of our heart, its tuning, and the relevance that emotions play in everything we do. Comparable to the story about the conductors in Zurich, the heart is also a conductor of electromagnetic impulses mainly generated by our emotions. If we can't *feel* it, we can't *heal* it.

The invitation is to enter in the realm of anahata, the sacred room of the heart where sound manifests. Stay awake. Allow music and sound to guide you from silence to possibilities as you explore vibration beyond thought. Therein lies mystery and the pulse of the universe.

CHAPTER 15

INTUITION

Choosing Bowls That Resonate with You

It is only with the heart that one can see rightly.
What is essential is invisible to the eye.

— **ANTOINE DE SAINT-EXUPÉRY**, *THE LITTLE PRINCE*

Guided by my intuition and the magnificent connection beyond the known with my son, my life was opening, and I was immersed in projects that were bringing great joy; teaching, playing events and concerts, interacting with sound colleagues, and creating new music. One day I received an e-mail that was to bring someone into my life who would soften my grief even more and bring greater focus to sound healing as a potent element in mainstream music. We share a love of singing and writing, a love of family, a dedication to bringing Light to the world through music, and we share the kind of loss that changes one's life path forever. She became the daughter I never had: my beloved daughter from another mother.

The message she sent me read:

Aloha. I am a singer/songwriter/mother from Los Angeles who has started to incorporate healing tones into my music, and sound baths and meditation into my live shows. I would love to come in and purchase some bowls, is this possible? I'd also like to sign up for the next sound healing certification class. I have been using Tibetan singing bowls and crystal singing bowls for personal healing for about 5 years now. I am so thankful for their healing. I am looking forward to hearing from you. Thank you for your time.

The e-mail included a photo of her three alchemy singing bowls: a grounded E note, a centering G♯ note, and an accelerating B note. Together these notes created a lovely E-major chord with the root note of E as the grounding tone. I learned later that E was one of her favorite keys to sing in. The alchemies of her bowls were:

- Laughing Buddha, made of iron oxide, which strengthens blood and activates joy
- Platinum, for balancing the emotional body and the chakra system
- Mother of Platinum for connection to the Divine Feminine

The e-mail was a normal communication the likes of which I receive regularly, although I was excited to learn this person was a singer and a mother. It is a privilege to consult and mentor other musicians and singers. When someone already has bowls and I am adding to their existing set or curating a new one, it is interesting to see which notes, tuning(s), and alchemies they have already chosen. This gives me good insight into the person and provides the information I need to enhance their set and create additional harmonics. I did not think any further about the e-mail, and we made an appointment for its writer to come to the studio.

A few days later, we met in person when this stunning young woman arrived with her equally beautiful sister at the Crystal Cadence Sound Healing Studio and Temple of Alchemy to explore the bowls. We immediately had a heart connection. I determined her three bowls were in the tuning of 432 Hz, and I proceeded to ask her questions and explore what the best combinations of notes and alchemies were for her. I gave the two women an overview of bio-energetics and the sublime world of the crystal singing bowls they had invited into their lives. They nodded their heads in understanding, and their eyes lit up with excitement as a new journey in music unfolded. They loved the combination of music, sound vibration, and science, as they were both singers and their father was a medical doctor.

"Each and every bowl has a 'consciousness,' meaning its own personality, and its own overtones and vibratory signature that are all determined by its alchemy, note, size, and tuning," I explained. "They are unique sonic beings just like we are. Often the first thing we do is determine which notes create a comfortable resonance and feel like a match for you. Normally people are drawn to certain notes or a particular key, and a tuning." Since my new client had brought her 432 Hz–tuned bowls, I taught the sisters the difference between the three primary tunings of 432 Hz, 440 Hz, and 528 Hz by giving examples of the singing bowls.

At any given time, there are more than 500 singing bowls available in the Crystal Cadence Sound Healing Studio and Temple of Alchemy, and an incredible vibrating energy is always present. During a personal consultation, we explore many aspects of crystal bowl sound healing to discover what resonates the most with you and who your "bowl mates" are. I may ask several questions including:

- What brought you to the alchemy singing bowls?
- Are you looking to add sound to your meditation practice?
- Are you wanting to add sound healing to an existing business?
- Are you inspired to integrate the singing bowls into your music?
- Do you know what musical notes you like?
- Do you know what tuning you want?

- Is there a particular area of your body or a chakra you are drawn to?
- Which crystals or which alchemies resonate with you?
- Are there particular colors you love?
- What is your comfortable budget?
- Is there something you are wanting to transform or heal?

The alchemies are minerals, gemstones, earth substances, and precious metals infused at high temperatures with the pure quartz, which add their own unique qualities to the bowl's resonance. For example, a client might choose between:

- Rose Quartz
- Amethyst
- Emerald
- Black Tourmaline
- Smokey Quartz
- Turquoise
- Moldavite
- Chrysoprase
- Gold
- Dead Sea Salt
- Charcoal
- And numerous other alchemies, which can be singular or in a blend of alchemies.

There is a synergy between the alchemy and the pure quartz, and in combination with the notes and the tunings, the sizes, and the shapes of the bowls, a unique vibro-acoustic experience is available to both the musician and the listener. It's fascinating how it all works together! Each bowl is handmade and one of a kind. What is the stabilizing magic resident in those combinations? Everyone uses them differently. My client base is broad in terms of their interests and businesses and includes those who wish to integrate sound in their personal lives, to professionals such as Reiki masters, psychotherapists, medical doctors, teachers, coaches, birth and death doulas, and yoga instructors who want to add sound healing to an established business model. I have curated sets for companies, health clubs, for lawyers, business consultants, schoolteachers, entrepreneurs, musicians, singers, composers, music school owners, psychologists, seniors, writers, athletes, students, inventors, and parents and families dealing with grief and loss. It gives me joy to help people integrate healing sound into their lives and the lives of their friends and family members. I love watching them land in a deeper happiness as they say, "yes" to their dreams and their visions through the power of intentional music.

Let's look at the consultation experience of the young woman I mentioned earlier, and her answers to my questions, to understand how people can choose a bowl to fit their life and their vibrational aspirations.

I began by asking my new client the question, "How did you come to the bowls?"

She replied, "My friend Krissy showed me someone playing them on social media. I saw they were sold in the city where I would soon be. I planned a day trip to the shop to buy my first bowl, and when I arrived and began to play them, I was astonished by how much I felt the crystalline sounds. Something unexpected opened in me, and I knew I wanted to keep exploring." She continued, "They're so very different from the Tibetan bowls that I have played for years and which I love. The sounds surprised me. . . " She searched for words to describe her first encounter with the bowls: "The vibrations were so beautiful and very calming for me—ethereal and celestial, but at the same time, they grounded me."

I understood her well, as I had loved and used the Tibetan bowls regularly too before finding the crystal bowls.

She shared that she was very interested in adding the alchemy bowls to her personal meditation practice, and we discussed new ways for her to do this. I then asked her if she would like to tone with the bowls using different vowels. She played a single bowl and toned. As she used her voice, I guided her to feel her breath flowing throughout her body, imagining it grounding her down to the earth, and opening her up to the radiance above her crown. We gently explored the chakras above and below the body, and she was fascinated. She felt how the sounds brought an anchoring and, simultaneously, an access to Light.

A NEW MUSICAL DIRECTION

As this young woman had mentioned in her e-mail, she told me her music was taking a new direction, and she wanted to integrate sound baths and meditations into her live performances. I was excited for her. As a musician and voice professor, it is my passion to support creative and artistic expansion. I had been integrating the bowls into my own performances already for a few years and knew the right microphones and acoustical setup for their optimal resonance. I was looking forward to helping her understand more about the possibilities of using her voice in combination with the alchemy singing bowls, integrating them naturally and with ease into her music.

She asked practical questions about the bowls and techniques of playing them, and she told me how excited she was to share them with her musicians. I asked her to sing a bit more for me, exploring different notes and different alchemies. She sang an improvised affirmation with each bowl. Her voice was stunning, shimmering in its timbre, luminous and unique in its qualities. It danced its way right into my heart.

I watched her joy grow as she exclaimed, "What I love is that these are easy for me to play while I'm singing. I don't feel like I have to rub my tummy and pat my head at the same time." She paused and demonstrated, and we all had a good laugh. "They inspire my songwriting," she continued. "They are not like learning to master the guitar or piano. I can't wait to explore writing and singing with them."

"Yes," I told her, "totally understood! And if I do my job right, which I do"—I laughed—"everything will always play harmonically and in tune."

This is not always the case with the bowls. I have helped many clients over the years who had spent thousands of dollars on sets of bowls they chose by their alchemy or how they looked but that did not match harmonically. If you are not a musician, it is hard to understand why the bowls don't sound well and how in combination with one another, they are not creating a safe container for healing. I explained to the sisters that when I first heard the bowls, I too simply fell in love with the exquisiteness of the sounds. My decision for my set had not been based on anything but the pure bliss I felt from them. This is true for most people. I had decided to purchase a chakra set, but at that time, I did not get to choose the individual bowls; they were chosen for me, and I did not see them until they were delivered. Like many people purchasing singing bowls, I did not have any specific information about them. It was a big jump for me to spend that much money on crystal bowls, but I was entranced by how the sounds made me feel.

"You bought an expensive crystal ball," my sister had joked. "Can it tell me my future?"

"No, it isn't a crystal *ball*," I said. "It's a crystal *bowl*!"

But it certainly ended up creating my future!

The women in my studio saw my passion, and I shared with them how much I had learned about the singing bowls over the past 17 years and how I now understood there were important factors that must not be ignored when choosing a set. Many combinations are possible, hence my commitment to guiding clients so that the bowls will be an amplification of their personal alchemy. We were on the same wavelength: they understood. The structure and integrity of music are everything.

We examined the notes my new client had already: the E, G♯, and B. I suggested F♯, A, and middle-octave E, which created sacred intervals. We played and we explored with notes and alchemies, always returning to what felt good for my client to sing with. Her creative flow had been ignited, and it was incredibly moving to watch it unfold.

Do you know what tuning you want?

In this case, she already had a 432 Hz–tuned chord, and my question was twofold. If she did not want to play the three-bowl set with other modern instruments, we could continue to build this set in the special tuning of A=432 Hz. If she wanted to play with piano, flute, harp, bass, etc., it would be preferred to take the standard tuning of music today: A=440 Hz. We expanded the 432 Hz set to be played on its own, which became the musical set for her first modern mantra single released in 2019, the "Trigger Protection Mantra." It was simply her 432 Hz–tuned bowls and her incredible voice, and it now has millions of streams.

Do you know what notes you prefer or what keys you like to sing in?

She knew she loved A and E notes, so we talked about what that meant from a sound-healing perspective, and we also discussed all the musical possibilities available when combining black and white notes.

I asked, "Would you like to play in the key of the heart chakra? Or the key of the throat chakra?"

Since my visitors were both in the field of music, they understood the question. I taught them the notes of the chakras in relation to musical keys and how that would influence a song. So, for example, if a song was in the key of F—the key of the heart—it would include one black note, the A♯, in combination with the white notes. They were fascinated by all the possibilities and clearly loved the key of E, the musical key of the solar plexus—the power of a thousand suns in the belly, self-esteem, and courage.

What is your comfortable budget?

The starting price of a Steinway grand piano is $70,000; a custom Fender Stratocaster guitar can cost $7,000 and up. A plain quartz bowl can cost as little as $200, but the price of an alchemy singing bowl begins at $700 and can go as high as $88,000 for hand-carved bowls. My new client gave me her budget, and we chose within that range.

Our conversation and sharing went deeper, and she expressed a bigger vision. "Thank you, Jeralyn, this is all very exciting," she said. "I see great expansion for myself and possibilities for many. I am so interested in helping people find balance, protection, and divine guidance through sound."

Then came the coherent thread that would tie it all together. The women asked about my journey, and I answered: Music since childhood, singing lessons at 11, a Broadway debut at a young age, followed by a new direction in classical music and a life in Europe: a fulfilling singing career, a wonderful child, a professorship, and the creation of my kids' foundation in Germany. I told them how I had discovered the bowls when my son was seven, our nightly ritual with them, his personal journey with the bowls, his subsequent passing at 19 and my life with music-recalibrating grief. They listened intently, tears flowing for us all. Then, vulnerably, my client and her sister shared that they had lost their beloved brother to brain cancer seven years earlier at the young age of 26. Our hearts broke wider open. I shared about my work with cancer patients, and my client told me about the song she had written for her brother.

We had been looking at a Moldavite bowl, which my client's sister loved. It was a grounded F#, an anchoring high heart note, and it was tuned to the music of today at 440 Hz. The Moldavite alchemy is incredibly powerful, creating a celestial bridge to other dimensions. She played it.

Her sister tried to find words for the bowl sound. "It's like a choir," she said. "There is so much depth, and it's such an ancient sound." I understood exactly what she was sensing and feeling. We all nodded about the importance of going deep with our pain and sadness and how music gives us space to process. Sound holds us and comforts us. My client added, "I've been in those dark places and every time, I've seen there is a brighter side. If I can just find the place of trust and open to the universe, I know

now it is all for a higher purpose. I'm a really sensitive person, I'm a Pisces," she smiled "and I know this to be true."

Our hearts connected, and we all felt a strong sense of knowing one another from a different time and place. We talked about communication through sound vibration and the signs of love we receive from the heavenly realm. We were bathing in the synchronicities and wonder of it all. We sat together in this place of sacredness, authentically sharing our real-life stories, bearing the sorrows of our hearts, and knowing how sound can truly be a medicine for our pain.

There was a poignant understanding that we were beginning a most miraculous journey of healing together accompanied by a new musical instrument filled with an unknown potential we all somehow trusted.

I curated an expanded set for my client based on the answers to the questions I had asked and what I experienced with her voice and her personal healing path. It was magical and Dylan was with me the entire time.

This wonderful lady then asked me if she could make an appointment and return with her manager, her team, and her musicians. She asked if I would please teach them about the bowls, the science, and the music theory, and explain it all to them as I had to her and her sister. Of course—my pleasure. I would love to do that! There is a powerful reason why I do what I do, and his name is Dylan. I love how the bowls bring joy and support transformation and healing. They had transmuted my grief and were doing wonders in the lives of many others I was teaching and sharing them with.

The next day my assistant was at the studio with me, going through our task list and e-mails, and as he saw this new client's e-mail, he asked me if it was "legit or a joke." I had no idea what he was talking about. "Not sure what you mean," I said. "She was here yesterday with her sister and chose a few bowls to add to her three-bowl set. We had a wonderful time together and have many things in common."

Intuition

He shook his head, raised his eyebrows, and looked at me, still unbelieving. "So, she was here?" he asked.

I nodded.

"Do you know who she is?" he asked.

I still did not understand. "Her name is Jhené, and her sister is Jamila," I said. "She is a mom and a singer looking to bring meditation and healing vibrations into her performances.

"Google her," he advised.

And that is how I met Jhené Aiko and her dear sister Mila J. A friendship guided by angels had begun that was to change both of our lives. The second time Jhené visited, her team came with her. I was planning the first live Sacred Science of Sound event in Hollywood at Wanderlust in the fall of 2019, a beautiful space with a long tradition of yoga and transformative work, a wonderful café, and a stage with great acoustics. We had engaged a handful of incredible presenters. I was excited for our first live event. I had set up a chakra set in C major for Jhené and her team to play with and then showed them the crystal alchemy beat frequency. As we were sharing and exploring more with the bowls, I heard Dylan whispering in my ear. He was fairly quiet at first, and then his voice became louder. "Mom," he said excitedly, "She is the voice of "modern mantra!"

"What? Son, please, I'm busy now. Let's talk about it later."

"No, Mom," he insisted, "ask her to perform at the live Sacred Science of Sound event. Play her some classic mantra. Play her the Gayatri Mantra and tell her what the Sanskrit words mean. She will be inspired to write a new genre of music called modern mantra. Discuss it with her and her manager *now!*"

I could not refuse his heavenly prodding.

Jhené loved the idea and resonated with what Dylan had shared. When he speaks as he did that day, his message cannot be ignored; it is expressing from higher consciousness. This is the bridge I traverse between Heaven and Earth that I so treasure. Jhené treasures that too. And so, our relationship has blossomed. We have since shared other live events, and I was honored to support the making of the three-time Grammy-nominated *Chilombo* album. Jhené's performance at the Sacred Science of Sound was a first for her, a sound healing for an audience that was totally unlike her normal public, followed by a conversation with me and a Q&A. The audience loved her and asked many great questions. This was the beginning of the unlikely pairing of a classical musician, professor, and singer who was doing pioneering work in crystalline sound healing, and an R&B artist—a singer-songwriter and author creating a new genre called modern mantra. We connected as part of a divine plan, through energy, frequency, sacred vibration, and our pain, held together by love and music. Jhené's modern mantra is healing for millions of people. "Alive and Well (Gratitude Mantra)" is one of my personal favorites. I am grateful to her and her whole family, to Mila J for accompanying her to our first meeting, and to their angel brother Miyagi Hasani, and of course, to Dylan. We all have a "bus stop" engagement together. It is a soulful friendship matched in Heaven.

I think Jhené best expressed her unique purpose and why she was guided to integrate the crystal singing bowls in her music when she said: "You know how a lot of people say, 'I lose myself in music,' or 'I like to escape'? I want my music to be more of an awakening. I want people to be aware of life; I don't want my music to be a distraction. I want to light a path."

CHAPTER 16

POSSIBILITIES

The Ability to Reach Beyond the Visible

Do you know how beautiful you are? I think not, my dear.
For as you talk of God,
I see great parades with wildly colorful bands
Streaming from your mind and heart,
Carrying wonderful and secret messages
To every corner of this world.

**—DANIEL LADINSKY, I HEARD GOD LAUGHING: POEMS
OF HOPE AND JOY: RENDERINGS OF HAFEZ**

It was a warm June evening, and the colors of the sky were filled with pastel light—an illuminated rainbow essence—and the air was soft. It was magical. I finished the last song, and the applause rang out. The audience stood up. I was grateful. I had just performed the *Forever Love* multimedia concert for the Science and Spirituality Conference in Nanaimo, Canada, alongside Gregg Braden, Dr. Joe Dispenza, Dr. Bruce Lipton, Lynn McTaggart, and Lee Carroll (Kryon). My musicians consisted of an orchestra of 24 bowls, which I played as I sang.

I had created a program of songs beginning with a classical aria, followed by excerpts from the *Forever Love* album. Every song was in a different musical key and infused with the exquisite vibrations of the singing bowls. On the large screen I shared images depicting communications from beyond the veil. The concert experience ended with a magnificent sound-bath meditation and a blessing to close the evening.

I am always surprised how every sound-healing concert is different. Being in the moment and fully present is so important. Bringing intentional sound to an audience involves first clearing and cleansing the space, infusing sacredness and harmony in the room, then humbly connecting with all those attending the event, known and unknown. It involves grounding and opening the heart to serve, aligning to the highest expression of Light and Love.

For this concert, I chose a magnificent set tuned to 528 Hz anchored with a C♯ 528 Hz–tuned Divine Kryon Bowl—for activating the embodiment of high-frequency Light. This alchemy was created especially for Lee Carroll and the channeling of Kryon. Some of the other alchemy singing bowls in this orchestra included a Rose Quartz A♯ to open the heart and activate self-love, a D♯ Platinum for emotional balance, and a G♯ Charcoal bowl for cleansing, clearing, and detoxification. I shared many of the stories I call "the divine miracles" with the audience: how Dylan's communication began in Los Angeles with the shooting star the night he crossed over, the beam of light on Dylan's picture as Dr. Sue held it atop Huayna Picchu, and the round rainbow containing the image of his face taken high above the Himalayas. I shared my story about the Dalai Lama, and Dylan's name on my car's screen playing "What a Wonderful World."

When I came down from the stage, there was a lineup of people who wanted to speak with me. It was all incredibly heartwarming, but one story stood out. A man in his early 30s, a Canadian farmer, said he had been going through some rough times.

He looked at me wide-eyed and said, "During the concert, your son came right to me like an arrow and began talking to me. I had been seriously contemplating taking my life. I don't know how else to tell you, as I am blown away. I felt his hand run down my face and sharply scratch my cheek with his nails as he said to me, 'No! Think of your mother, how much she loves you. You have much to do with your life. This is not your time.'"

This young man was so moved, as was I. He was in awe from the whole experience, especially the physical sensation of having his cheek scratched. I have never forgotten him nor the concert and the miracles it brought.

THE MAN IN THE ROOM

There was another experience with a cancer patient whom I will never forget either. She jumped up excitedly following the sound bath meditation and exclaimed:

"Who was that stranger, that man in the room?"

I was hosting a sound bath at the Cancer Support Community. The room was full. I began with an invocation as I always do, setting a sacred space and asking everyone to set their personal intention. I led a guided meditation and then played the bowls in silence. I was about 20 minutes into the experience when I suddenly found myself toning in a very particular way. My throat was open, and the tones emerging from me were the lowest tones I've ever made—almost like growling with a strange combination of consonants. At times it sounded like words, but if it was a language, I had no idea what it was. It was simply flowing out of me.

I kept playing while these sounds resonated out of my whole being. The tones were toning me, and I became the instrument for their expression until they stopped. I then closed the session as I always do with the gentle sounds of the chimes. As I tenderly opened up the space for listeners to share their experiences, someone shouted, breaking the pristine silence in the room, "Who was that man?" This caught me by surprise. Someone kindly responded that there was no man there; it was just Jeralyn singing in her low voice. The shouter responded insistently: "No! there was an Indian man here. He took me by the hand, and we boarded a train!" I was astonished, speechless. She continued with great conviction: "This just happened! We had a splendid and miraculous journey. He healed me! Who was he? And where is he?"

And then: "I saw him too!" "Me too!" "Who was he?" three other voices exclaimed.

I came home from this experience and sat in silence, wondering what had happened. My mind was questioning, but I was in grace. I had never experienced anything like this. It took me a while to wrap my head around it, to fully integrate and recalibrate myself. Although I cannot logically explain this, all I can say is that after this day, my sound baths were never the same.

THE ALCHEMY OF ACCEPTANCE

I have been curating singing bowl sets for clients around the world since 2016. Every experience with my clients is unique and fulfilling. Brenda's story, however, is truly astonishing.

Brenda is a therapist I met while I was performing at an event. She also leads retreats for women, and wanted guidance to integrate sound healing into her practice.

We met online for our Zoom appointment early in December 2021. She was seated in front of her Christmas tree, which was decorated with beautiful turquoise and purple balls and ribbons. I began asking Brenda the questions I normally ask to understand her needs and see which bowls would be best for her.

I was suddenly interrupted by my heavenly helper's voice. I heard it loud and clear: "Mom, go get her the purple bowl and build her set around it!"

There was no ignoring him.

But I responded, "No! I object. I love that bowl, and it stays with me."

"No way, Mom," he said. "Go and get her that purple bowl and build the set around it." By then I knew better than to continue arguing with him, so I went to get the bowl. Of course, Brenda loved it. It was a 432 Hz–tuned root-chakra C note with the alchemies of Violet Aura Gold and Apophyllite. It was exactly the color of her purple Christmas decorations. The set that was coming together was absolutely perfect for Brenda, and she was ecstatic!

Not happy with just one of my personal bowls, then Dylan told me to add my Divine Kryon bowl—a very rare small one, by the way. "No, son," I said. "That is mine."

"Yes, Mom, it was yours. Now it's Brenda's. Go get it! No bowl hoarding here!" he commanded as he laughed. I laughed too, recognizing that my heavenly assistant is sometimes my boss!

We built her a gorgeous set with seven bowls based on a deeply grounded F bowl, creating a mini crystal orchestra in the key of F, the key of the heart. Brenda loved her set and trained with me in Level 1 of the Sacred Science of Sound. She played sound baths for her clients who adored them.

A few weeks before her Level 2 training was to begin, I received a call. Brenda was sobbing; she could barely speak. "Jeralyn, I was preparing to lead a retreat. My bowls were set on a folding table when I heard a big crash. I ran back to the room where the bowls were. The table had collapsed, shattering all but two of my bowls—which were perched unharmed on top of the quartz rubble! My set is in pieces."

I sensed it as she told me, although I found it unbelievable; the two bowls Dylan had chosen for her: the purple bowl and the Divine Kryon bowl, were the ones that remained intact, perched atop the pile of shattered quartz. I listened in wonder.

After the shock wore off, Brenda recognized the bigger picture message as to why this happened, and asked me to rebuild her set based on the two surviving bowls. And what came together was clearly a more complex orchestra.

What does it signify when a bowl breaks? For Brenda, it opened up an understanding of what she needed to let go of to grow and evolve. The breaking of her bowls created a huge energetic shift for her. She was never the same: transformed by loss and reinvented through acceptance. And it was not a coincidence that her new set included more alchemies of the heart, such as Rose Quartz, Rhodochrosite, and Emerald.

HACKED BY SPIRIT

In May 2020, I was invited to participate as a speaker in the second Minerva University graduation, which was being held online due to Covid. As the ceremony began, I could not help but think of Dylan and how much my boy would have loved attending this university. Within the first few minutes, my computer screen went black. As I sat gazing at the dark screen, I froze. My initial thought was, "Is my computer being hacked?" A friend had recently been hacked, and this was on my mind.

The moment I noticed my head making up a story and taking me for a ride, I took a deep breath to calm my mind. There was no one I could call to help me, and all I could do was to accept the situation and wait. As I inhaled, my body softened, and the next thing I knew, the screen of the computer was coming back to life. But for some strange reason, it restarted opening my Skype app, which I never use. I looked at the screen with surprise, and I read: "Dylan Sage, last seen days ago."

I held my breath. I had in front of my eyes a message Dylan had sent me more than six years before.

"Hey, Mom, I'm here. I Love you," his message said. "All is good."

I was indeed being hacked. But not in the way I had thought!

LIFE-CHANGING TOOLS

I had already created a powerful 432 Hz-bowl set with the notes of the chakras for one of my students in Europe, who is also a Reiki master and yoga teacher. She was having so much positive feedback from her sound baths that she asked me to curate a second set for her, a totally different one, in the tuning of 528 Hz and in pinks and greens—colors of the heart. I built the set over a number of months with Rose Quartz, Rhodochrosite, Emerald, Peridot, Pink Aura Gold, and Saint Germain. Sometimes putting a set together with the specific notes, tunings, sizes, and alchemies, plus having them nest safely together in a case, is like searching for a needle in a haystack! I don't give up, because I know these are tools for a lifetime. In some instances, it's taken over a year to find just the right bowl to finish a set perfectly.

My student Laura Israel from Hamburg, Germany, was thrilled with her new instruments and the feedback she was getting. "The bowls are soooo harmonic, the perfect mix of deep and high tones. People are telling me, 'It's flawless, like the set had to be exactly that combination!' It's also incredibly calming for the nervous system."

One morning, however, she called me expressing concern. "Have I done something wrong?" she asked. "Could you please explain the reaction two of my clients had with my new set? As I was playing the low-D♯ Rose Quartz bowl, one woman experienced great discomfort and stress in her body. She wanted to get up and leave. The other woman felt intense pain in her lower belly."

I first reminded her that sound vibration may trigger uncomfortable feelings, physical or emotional, in both the receiver and the practitioner. Then I encouraged her to remember to use the tools she was taught to create a safe space and trust herself. I said, "Next time, you could invite people to notice where in their bodies the discomfort sits and then guide them to breathe with intention into those places and feel. Also, toning with intention in those places can be very helpful. This may allow people to release tension and support integration of scattered aspects of themselves. When someone is feeling discomfort, you can remind them that all feelings and emotions are simply energy that can transform. This may help calm the mind and ease the body."

Our next call was very different. She was happy to share that after implementing these concepts, her two clients were able to breathe, feel, and accept the discomfort and move through it with toning. "I am so grateful! These bowls and your guidance have changed my life!" she exclaimed.

BOWLS AND HEARING CHALLENGES

I am often asked if people with hearing challenges, such as deafness and tinnitus, can enjoy the bowls. The answer is absolutely, yes! A few years ago, I was honored to open an event with journalist Maria Shriver. She asked me to set a tone that would de-stress her guests and bring them focus and clarity. I chose to begin the evening with a set based on the pentatonic scale. After I finished playing, people asked questions: they had never experienced the singing bowls and were very moved. One woman stood up and shared how astonished she was to find that following the short meditative sound bath I had delivered, the tinnitus she had suffered from for the previous 10 years was suddenly gone. I have also worked with other clients whose tinnitus disappeared after a few sessions. Sound can truly and inexplicably tune and reshape us.

One of our students, who worked with deaf people, would put her bowl on their bodies so they could sense the sound through vibrations. Regarding hearing aids and sound baths, I will always defer to the fact that sound must feel comfortable, which is something I really learned through my mother. She has hearing enhancements, and at times she has told me a particular sound hurt her ears. I find this so interesting, as it has happened with different sets and different combinations of notes. Sometimes it will happen with higher notes and sometimes when I am playing lower notes. With time I have found a gentler, simpler way to combine notes and wand techniques geared exactly for my mom. This underscores how important it is for the person playing the bowls to be sensitive to the individual needs of the person or group receiving the sound bath. This is why it is crucial, if you are giving a group sound bath, to feel the group's energy and resonance and respond accordingly. If you are receiving a sound bath and feel discomfort, breathe into the area of your body that is activated, and see if your breath helps the discomfort to dissapate. If it does not subside, please let the practitioner know.

If the sound is too uncomfortable, it may not be possible to relax. This can be the case for people with hearing aids, which amplify frequencies. Sometimes people with removable hearing aids ask if they should take them off during a sound bath. If they do, they may choose to sit as close to the bowls as possible and receive a different experience of hearing and feeling the sound. It is a personal preference. If someone cannot remove their hearing aids, I suggest they sit farther away from the bowls to avoid any discomfort that may occur.

There are times where unpleasant sound becomes a gateway to some revelation. I am reminded that within my own intense healing process, I experienced the nails-on-the-chalkboard sound as my grief was at its height, and in time recognized it as a reflection of my own pain as it was transmuting. I worked through it with breath, embracing the despair until it eased and the sound became coherent. The alchemy bowls work with a higher intelligence (one beyond thought), and the most important element lies in playing and receiving with love.

LAUREN'S STORY

My passion for music and my commitment to excellence has led me to meet and perform for world leaders and many successful people in a diversity of fields. One of them is the beautiful actress Lauren London, a pillar of courage and strength who lights up the world with her radiance. "All you have is you and the Divine," she says.

Lauren grew up with a strong connection to God to hold her steady throughout a turbulent childhood. But when her beloved partner, the admired American rapper, philanthropist, entrepreneur, and activist Nipsey Hussle, serving through his music and committed action, was killed by a bullet, she was called to a higher purpose: as she states, "to survive and thrive through the grace of God." Life taught her the importance of being able to sit in pain, knowing that God had her back. So when the father of her child passed away from gun violence, she knew she had to sit in the pain to transform it. With a deeper understanding of the flow of life came trust. As she turned her pain into a focused purpose, music and her faith accompanied her along the way.

"It's not easy," she says, "but I am here to serve. No pain is in vain, and I am transmuting pain into purpose. Step inside of yourself when you are in pain, and step outside to serve."

Which is what she did. Which is what I instinctively knew to do also, by volunteering to work with cancer patients and veterans not so long after my son's passing. I met Lauren at a celebration where I was invited to play with beloved friend and exquisite devotional singer Jahnavi Harrison and keyboardist-arranger Chris Sholar. The combination of Chris's smooth, fluid improvisation, Jahnavi's rich, expressive voice sharing sacred mantra, and the crystal singing bowls was sublime. Most of the guests had never experienced a live sound-healing concert before, and they were amazed at what they felt throughout the experience. I had a few different sets of some of my favorite bowls with me, including three large Supergrade bowls that played a magical, low-octave beat frequency.

What is it about these deep sounds, representative of the male voice range, that is so compelling? For me, it confers an unwavering sense of safety and protection. This type of sound allows for experiences of expanded awareness and a kind of exploration that can go beyond time and space. This is what I had in mind when I chose the bowls for the event. When I finished playing, I was curious to know what the sound had evoked in the guests, so I invited those in the audience to share their experiences. Some people spoke about feelings of inner peace and calm. Others spoke about positive energy and renewal, while still others felt the safety to release heaviness, pain, and even jealousy. Many said it was so deep, they had no words at all.

Almost a year later, Lauren surprised me when she shared with me that she had seen a small blond boy around me during that sound-healing concert. He was being bathed in water, and he was smiling. She knew it was Dylan, even though she did not know what he looked like. She also reminded me that he is always with me. She is raising her two sons with values she holds dear to her heart: integrity, honesty, truthfulness—"pillars of a higher consciousness," she calls them. Her older son experienced his first sound bath with me that put him into a deeply regenerative sleep. When it was time to go home Lauren gently awakened him, and he shared that the sound bath was "chill." He loved the science I had explained and found it all super cool.

Lauren and I share the unexpected loss of a loved one way before their time. We have spoken about the power of music and how the sounds of the singing bowls bring a sense of stability, focus, and clarity, creating a sense of safety that helps us release frustration and anger or anything that no longer serves us. I love Lauren's radiance and great sense of humor. She always makes me laugh and brings me joy. She has told me often that I bring joy into her life and that my music is both a physical and emotional medicine for her. For this I am truly grateful.

SEAN'S STORY: BRINGING THE HEALING TO SOUND HEALING

I began playing regular sound baths in Los Angeles, curating different combinations of singing bowl sets selected from the large collection of instruments in the Crystal Cadence Sound Healing Studio and Temple of Alchemy.

I had been sharing healing music for about three years when in April 2019, I played a sound bath I will always remember. I had created a powerful set utilizing the sacred intervals to ground and tune the heart. I have come to trust that the unique sets that I build for any given event always fit the energy needed. This set was in the key of F major, the musical key corresponding to the heart, and I felt this tonality was important to use on this night. I chose the alchemies accordingly; among them were Lemurian Seed, Morganite, Pink Tourmaline, Rose Quartz, and Andara. The notes of this set intentionally evoked the connection Heaven-Earth: inviting the ascending energy of the Legacy chakra to rise up through the body and reach the heart to meet the descending energy of the Life Purpose chakra. This exchange creates a stable circuit that anchors us to the earth while simultaneously opening a connection to the Divine.

It was a balmy night; the room was packed, beautifully lit to encourage a deep sanctuary of meditation, and there were a few faces I had never seen before. Jhené Aiko was present and had brought a guest whom I had not yet met. Both were seeking renewal, relaxation, and healing. Normally before a sound bath, the room is alive with expectation, and after a sound bath, the room is filled with wonder and reverie. People enjoy the afterglow of the crystalline sound; they are mesmerized, and no one wants to leave. I often hear, "We don't want to go home! Can we stay and have a sound bath slumber party?" That feedback always makes me smile.

In the sacred space we co-created that evening, I invited the guests to speak into their experience. The authentic sharing supports everyone to integrate, embody, and ground. One of the new participants spoke openly: "It's been a stressful week," he said. "The sound bath actually soothed me. How can I help people to understand that certain actions have consequences that can't be undone? Like acting out and creating permanent solutions to temporary problems?" He spoke to the importance of having compassion and tolerance and being less judgmental of one another.

The sound bath had created a circle of trust, safety, and intimacy for him to allow his heart to speak.

The room was still as he continued: "The sound calmed me, and I felt a deep gratitude for being alive. I've been to sound baths before, but never one like this with the alchemy singing bowls. It is the first time I actually experienced the healing aspect of sound healing: I felt somehow the weight of the world was lifted off my shoulders. I felt lighter, like I do when I meditate—similar, but a different kind of experience: relaxing. I felt totally supported. I was so disappointed that my friend died—for nothing—and this experience of sound. . . for me, it was for real; I felt the healing.

"The pressure of life is intense right now with everything going on in the world, plus the stress of my own busy schedule; finding my balance and learning to live in the present moment alongside the expectations, relationships, and the overwhelm. I recognized with my friend's death how important it is to be with those you care about and not put things off because your habit is to prioritize responsibilities and work. I realized I count on people always being there physically and I've been feeling some guilt and regret. Me and my friend Nipsey were making a song together that wasn't yet finished. . . and then he passed away. I finished the song 'Deep Reverence' to honor him. It was one of my proudest career moments to share that song. Although it wasn't our first recording together, it was our first song with only the two of us. It was especially meaningful because it was one of the last verses Nipsey recorded. Respect people's presence in your life and don't take anyone for granted." ("Deep Reverence" later received a Grammy nomination.)

The conversation was deep and it was real. I shared about Dylan's death, and the intimacy of the space opened even more. The man was moved. He said that hearing me speak reminded him that we never know what people are really going through just by looking at them. He emphasized, "We must be understanding and kind at all times and avoid being judgmental. This is the only way to get to any progression within yourself and for the planet to evolve."

He nodded his head in realization. "Music as medicine can go deep. I shed tears during the sound bath, and I hugged Jhené," he said, acknowledging the woman beside him on the yoga mat. "I felt the frequency of Home and so much gratitude for being alive, for feeling, for loving. I understood the bigger arena we are all playing in. I'm grateful for this pure energy of sound. I recognized no matter what, everything is always gonna be fine. This belief is imbedded in me now."

As the three of us walked out of the event, he shared what it had been like to grow up in Detroit. He spoke about his family and his passion for music. I asked him more about his recordings, and he suggested I listen to "One Man Can Change the World" and formally introduced himself. "I'm Big Sean," he said. And his friend was Nipsey Hussle, actress Lauren London's beloved partner.

We stood outside and talked for a while longer. Sean shared that the moments of despair he had experienced in his life helped him to believe even more. "They have given me faith," he said. He recalled attending the Church of Today outside Detroit as a young child, and at that time, Marianne Williamson was the pastor. I told him that I had recently played for her Los Angeles event and, by her request, began the program singing "Amazing Grace" and playing the singing bowls. Sean shared that he loved when I sang the song at the end of the sound bath, and it was an unexpected, touching closing for him, bringing pure medicine for the spirit.

Music cannot change situations or life circumstances, but it can help us shift our emotional response to them, bringing a sense of safety, comfort, and strength, allowing healing to happen. As we ground, we rise, learning to trust in the elevated view of life, accepting both the pain and the beauty of it all.

LOVE DRIVES THE BUS

Most of us are conditioned to push away what we don't want to feel and what we deem to be "bad." However, with or without the crystal bowls, it is essential to remember that the main transformer is the heart, and the heart is electric. It is the essential connector and must be tuned to the frequency of love. By accepting what life brings, the body softens, stops fighting and flighting, and we transform. Acceptance is a frequency that is in harmony with all that is, an invitation to be present in the here and now. Right in this moment, you have the power to decide to accept yourself, your feelings, and your circumstances exactly as they are. Notice what happens when you allow this frequency to ignite within you.

Accepting my life as it now was when it so suddenly and painfully shifted is what allowed the love within me to reveal itself. I stopped asking why. I stopped trying to make logical sense. When I accepted the excruciating feeling of loss and had the courage to be present with it rather than begging it to go away, I gave permission for Love to come in and take over. It has been "driving the bus" ever since.

CHAPTER 17

VISION

Collaboration, Innovation, and the Future of Sound Medicine

> Hope is the thing with feathers
> That perches in the soul
> And sings the tune without the words
> And never stops at all. . .
>
> **— EMILY DICKINSON, NO. 254, STANZA 1**

> When we quit thinking primarily about ourselves and our own self-preservation, we undergo a truly heroic transformation of consciousness.
>
> **— JOSEPH CAMPBELL, *THE POWER OF MYTH***

"Instant everything" seems to be one of today's main themes: instant solutions, instant relief, instant rewards, instant relationships, instant success, instant popularity, even instant answers with AI. More than ever, our world has become a very fast-paced and outwardly focused soul grinder.

Is it perhaps time to stop and listen to the Music in a new way? This is my hope for today. This is my hope for our future.

I CALL ON HOPE

In August 2023, I experienced firsthand the catastrophic fires in Maui. It reminded me just how valuable music medicine is becoming, especially to soothe us in situations we cannot change. I asked myself what I could do as a musician and sound therapist to bring relief to a population in shock, loss, and grief. I prayed. Witnessing the immensity of the devastation brought me to the brutal realization that nothing, not even time, could bring back what had been lost in an instant. All I could do was to accept, to breathe, and do what I do. I played crystalline music and sang. I set a powerful intention of healing, held sacred space, and sent my love to Hawaii, to her land and her people.

It is time to cultivate and amplify the essence of Music in a bigger way than we have ever done: we need the succession of notes, of harmonies, of words, of melodies, of frequencies that create a coherent whole and weave us together in unity. Whether we find it in the power of a simple refrain or a complex masterpiece, we commit to music to be an integral part of our lives, deeper and more expansive than ever before. We carry on, we persist on the path of sacred sound vibration. And we understand that in any great composition, be it a song we love or a symphony, music has a progression, its own unrushed journey, characterized by its rhythm, tempo, and structure. Improvised music unfolds in the present moment and creates a powerful expression; however, most music is not instant and takes time to develop and evolve. Music brings us to unexpected places. It is the expression of hope being planted. Growing roots. Transforming us. Now.

I cherish my relationships with my colleagues, many of whom I have the privilege to call friends. It is inspiring when we can collaborate, and these collaborations are gifts that keep nurturing. Music teaches us how to work together in harmony. This support lightens the load we are all lifting. I encourage you to collaborate. The connection and possibilities are invaluable and unlimited. There is Trust. Integrity. Grounding. Compassion and Kindness. Like a great orchestra, we tune, we are poised and ready, and we await our downbeat. We play with dedication, passion, and in service. And what "sound" seeds are we planting now? We are creating a rap song, and it goes like this: Unification. Participation. Communication. Meditation. Conversation. Integration. Innovation. Revelation. Transformation. Inspiration. Education. Music, the prescription-free Medication.

VICTOR WOOTEN: MASTER MUSICIAN AND INNOVATOR

Victor Wooten is one of the greatest bass players of all time. He is also the author of two books and leads Vix Music Camps, where students from around the world study music and nature. I am grateful to call him a friend, and I love our inspired conversations.

"Music has a positive effect on people of all ages," he says. "It is a medium that brings people together and often assists us in many parts of our lives. Music has helped people to walk, talk, think, meditate, dream, relax, exercise, celebrate, grieve, heal, and more. It has even helped bring unconscious people back to Consciousness. When Music is present, we don't care who voted for whom, to whom someone prays, how much money someone makes, or the color of someone's skin. Music brings us together without force or prejudice. That is powerful!

"Studies have shown that Music helps children do better academically. It helps them work together, listen, and exist in harmony. Children have always responded to Music. They sing, dance, and play without inhibition. They approach Music naturally as if it is easy. But still, Music is underfunded and steadily being removed from school curricula. With all that being said, I am still convinced that Music's future is in good hands.

"Currently, scientists are beginning to recognize Music's power and are studying it in a new way. They want to know how us musicians do what we do. It's as if they know that we are superheroes. That really excites me. Although scientists are mostly discovering what us musicians and our ancestors have known for centuries, their 'proof' may bring attention that will hopefully result in much-needed funding.

"Music as medicine is a steadily growing subject. Music therapy can even be studied in college. Although I am super happy about it, I know that we are just scratching the surface. As a society, we have forgotten what our ancestors have known since the beginning of time. But we are waking up. We are remembering and recognizing Music's importance.

"Because you are reading these words, you are my proof that we are moving in a good direction. Thank you for your time, attention, and awareness. Together we will remind all of society that Music is powerful, important, and necessary for life to exist in harmony. I look forward to our paths crossing as our musical journeys continue."

DR. DANIEL J. LEVITIN: SOUND THERAPY AVAILABLE TO ALL

Recently one of our original collaborators with the Sacred Science of Sound, neuroscientist Dr. Levitin, shared with me what is developing in the field of music medicine. I so appreciate his presence and guidance in our work.

"Licensed MDs are prescribing it," says Dr. Levitin. "Hospitals are implementing music medicine programs; health-insurance companies and Medicare now have reimbursement codes for it." Yes! This has begun and is happening now. And as we learned at the Conference for Music as Medicine in Washington, the National Institutes of Health has funding to support some of these studies—and is doing so.

Dr. Levitin began his career as a musician and music producer before he became an eminent author, researcher, and educator in the field of neuroscience, where he is especially notable for his pioneering work in the area of music and the brain.

He says that he believes we are moving toward a world where music will become a priority and be seen as a necessity. From preschools, elementary, middle, and high schools to universities, all education will integrate music into its program offerings. Innovative music programs will make sound therapy available to all students to support their mental health, academic success, focus, learning, emotional stability, and sense of belonging.

In 2017, Dr. Levitin attended my concert premiere of the album *Forever Love*, which included spoken meditation and songs scored with crystal singing bowls. He shared with me that both he and his wife were very touched by the uncommon beauty and the amplitude of their sounds, in combination with my vocals. He had never heard crystal singing bowls before, much less heard them integrated into songs with other instruments. He said it helped him appreciate the singing bowls as an important tool for health and well-being, and now seven years later, he is looking forward to what will be revealed as the newly formed Minerva Lab pioneers research that will include the crystal singing bowls.

THE MINERVA UNIVERSITY LABORATORY FOR THE SCIENCE OF MUSIC, HEALTH, AND WELLNESS BEGINS!

The Laboratory for the Science of Music, Health, and Wellness at Minerva University began its first undergraduate fellowship program in the summer of 2023, under the guidance of Dr. Levitin and supported by distinguished Minerva University faculty including Mark Sheskin (Ph.D. in psychology, Yale University); Katie McAllister (Ph.D. in neuroscience, Cambridge University); and Randi Doyle (Ph.D. in psychology, University of New Brunswick).

The Lab's mission is to elucidate the science underlying the ancient connection between music and medicine and, more generally, between sound and health. A critical component of this work is to train the next generation of top scientists, engineers, innovators, researchers, and practitioners. Using the latest techniques, the Lab aims to establish the underlying mechanisms by which sound therapies work, to promote the findings in the widest possible way, to set the standard by which other laboratories operate, and to influence research around the world.

The Lab supports research and start-up incubation projects on the impact of music and sound on health and well-being, and it currently supports an important research partnership with Dolby

Labs. What's particularly exciting about Minerva University is that it's the most globally diverse university in the world—with students from 100 nations and campuses around the globe! This means it has a unique opportunity to embed a cross-cultural element in all research undertaken.

At Minerva, undergraduate and graduate students—many of whom are entrepreneurs—work in close consultation with faculty mentors. Their training involves not just scientific pursuits, but also the effective communication of their findings, both to other scientists and to the public. The Lab brings together expertise from disciplines that have historically been separated, including music, psychology, neuroscience, physics, computer science, and engineering. It builds on deep expertise within Minerva University in the areas of neuroscience, psychology, music, and artificial intelligence.

The training also involves the development of partnerships with industry to better prepare students for the workplace. Binaural beats, brain health, and crystal singing bowls are a part of these studies, as are other instruments and methods related to sound and health. The plan is for the Lab to grow dramatically over the next several years, and it promises to play an important role in shaping the future of sound medicine.

JAHNAVI HARRISON: RECONNECTING TO THE SACRED SPACE WITHIN

Devotional singer and musician of great humility whom I accompanied on her album *Balm*, Jahnavi Harrison brings a timeless grace and beauty with her soulful voice, touching us deeply and opening our hearts. I am honored to share with you her vision of music.

She says, "I believe that sound vibration, especially music and singing in community with others, are some of the most potent medicines to heal body, mind, and spirit. For myself, the practice of devotional chanting of the Divine names [*kirtan*] has been in my life since the beginning and is the primary source of my experience and faith in the healing power of sound. Just as in recent decades modern healthcare has given greater acknowledgment to indigenous healing modalities, I envision that over time, the power of music and singing to affect deep change can be explored, facilitated, and experienced globally.

"One of the most poignant experiences I had in recent years was during the first months of the Covid-19 lockdowns. I felt the need to establish a daily connection to prayer and chanting for my own mind and heart. I had already been singing on Zoom for several in my community who were sick, and it occurred to me to broadcast the daily devotional chanting online so that everyone at home could join in virtually. I can't recall ever having conducted kirtan online in that way before, and I certainly had no idea how many people would end up joining. As the days went by, more and more people attended the daily sessions. Most days, about 700 to 1,000 people were online at the same time, chanting or connecting with the sacred sound vibration from home. Though all I could see were names or sometimes just numbers on the screen, I often got goose bumps as we chanted, feeling that I could 'hear' people's voices calling in response. It was one of the most beautiful experiences of my life. We continued daily for six months and then once a week for another year.

"Now as I travel, I very often meet people who joined in online at that time, and I'm astonished to hear their stories of how it helped them. I knew it was helping me. . . and to now know it was helping others so deeply too is a great blessing. When I held my first-ever tour in 2023, I finally got to sing in the same room with many of those individuals—sometimes as many as 1,400 voices together. It was overwhelming in the best way. Of course, relatively speaking, it's a small offering, but I think that is also the beautiful thing about it.

"I am honored and grateful to have had the chance to offer some tiny droplets to the wide, fathomless ocean of sacred sound vibration, and I believe every person with a voice can also be empowered to use it in a way that offers light and healing to others, even if just within our circle of family or friends. For myself, no matter what new surprises are ahead on the journey, I hope I can be in service of sacred sound as long as I am living."

DAYVIN HALLMON: MUSICIANS AS FIRST RESPONDERS

A shining example of a sonic activist, a soundworker on the front lines, is Dayvin Hallmon. Dayvin is a multitalented musician who was called to do more in Milwaukee, Wisconsin, where he currently lives, and so he created a string ensemble of Black and Latinx musicians to go into communities where violence has occurred and play music. PBS has created an inspiring 12-minute film entitled *When Musicians Become First Responders* that shows what a powerful medicine music is! For those who have experienced tragedy or violence firsthand, Dayvin's goal is to ensure that the seed of destruction that these experiences plant does not take root and play repeatedly in people's minds. The "first responder" musicians help soothe the shock and move those affected immediately across that river of despair which leads to hopelessness. His project is truly inspiring and he and his team are out in the field where tragedy lives, bringing real healing through their music.

The mission statement of the Black String Triage Ensemble eloquently conveys its musical role within the community: "Using Black Music and Art to address pain; foster healing; promote love; call for justice; and guard against hopelessness."[1] Guarding against hopelessness is a medicine we need, especially in our inner-city communities.

JOHN STUART REID: THE FUTURE OF SOUND AND MUSIC-BASED MEDICINE

You met John Stuart Reid, an acoustic-physics scientist and authority in the field of cymatics (the science of visible sound) earlier in this book. My conversations with him are equally inspiring, impassioned, and educational for me.

John shares with us that a discovery Professor James Gimzewski of UCLA made in 2002 offers an intriguing potential for eradicating not only cancer cells, but perhaps any pathogen as well.[2]

He says, "Using an atomic-force microscope, he, his colleague Dr. Andrew Pelling, and team were able to listen to the sounds of cells for the first time. Surprisingly, they found that the respiration sounds of cells lie in the audible range when amplified, and so they named their new approach to cell biology *sonocytology*, referring to the songs' of cells."

"This leads to the exciting possibility of a future in which we could see the drug-free eradication of cancer cells. By taking a biopsy of a cancer, its sonic signature could be detected and amplified, and then used to modulate an ultrasound beam directed at a tumor. In such a scenario, the tumor cells would absorb sufficient acoustic energy (of the cancer cell's own sonic signature) to be destroyed. Such a therapeutic procedure would likely be given during a series of outpatient visits, in which a percentage of the tumor's mass would undergo a controlled shrink during each session to minimize the toxic waste of dead cancer-cell material. For leukemia sufferers, this principle holds the potential for sonic irradiation of the patient's blood via a specially adapted intraoperative recirculating system.

"In another study I conducted with Professor [Sungchul] Ji (professor emeritus of Theoretical Cell Biology at Rutgers University), which was published in the research journal *Water*, sounds from cancer cells and healthy cells were made visible with the aid of a CymaScope instrument, imprinting the sound vibrations onto medical-grade water—rather like a fingerprint on glass—thus leaving a visual signature of the cell sounds. A typical CymaGlyph [sound image] of a healthy cell sound is symmetrical, while that of a cancer cell is skewed by comparison. This collaborative study was a first step toward developing a sonic-based technology with the potential to eradicate all cancer cells, whether in tumors or in blood."[3]

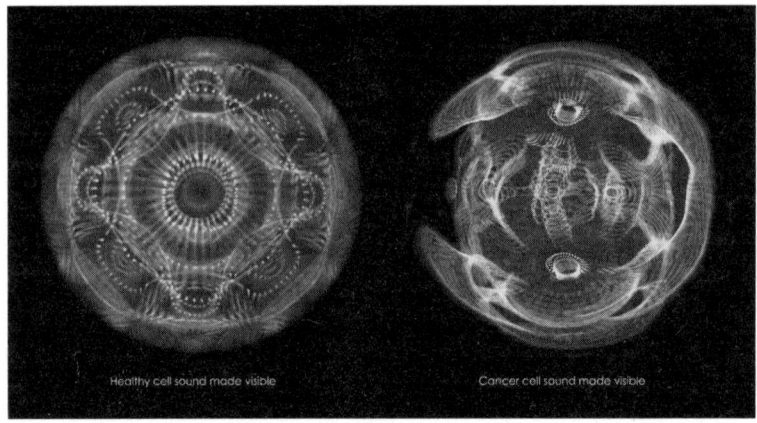

Healthy cell CymaGlyph (left), cancer cell CymaGlyph (right).

There are exciting possibilities emerging in the collaboration of music as medicine and cancer research.

KEVIN JAMES: HEART SONGS: KEEPING A HEALTHY, HARMONIOUS CULTURE ALIVE

Kevin James is a spiritual musician from Australia whose songs are a moving combination of sacred mantra and original material. Kevin's Heartsongs Retreats, which include practices, teachings, singing together, and days filled with musical offerings, are transforming lives around the world. His music touches the heart and soul. I am honored to share his voice here. As you read in Chapter 8, his music was medicine in my life and has held me safe and comforted on the road of grief.

Kevin says, "My vision for the future of healing sound is the same one I've had since the beginning of my journey with music.

"Our music is a contribution to the culture of connection, harmony, and healing—transcending the limitations of language and bypassing the analytical thinking mind—to assist anyone around the world who resonates and is willing to align with their true nature or essence.

"We carry this flame in our music and the way we live. Like the heart of the sun, we hold it dear and hold it near to our own heart, and it is our joy and privilege to warm the hearts of others with our song.

"For me, healing and harmonious music is a guiding light into the heart of a healthy culture of connection because it can transform our consciousness and thus the way we live; and it is culture that will continue long after we're gone.

"I believe that it's very important to keep a healthy, harmonious culture alive through music and sacred sound, especially now in these times of change."

JOSHUA LEEDS: "ONWARDS TO THE RENAISSANCE"

I love the deep conversations I have with Joshua Leeds, an esteemed music producer, sound researcher, and author. His books include *The Power of Sound* and *Soundwork on a Hot Rock*. I'm honored to share this very personal piece of writing from Joshua about the future:

> While I live in a quiet mountain cabin, I know that in most places, there's dissonant noise deriving from traumatic sadness. We're witnessing a lot of *global* changes, where many are anxious, across borders, at the same time. Humans have lived too hard on Earth's ecosystem, and the planet is reacting with sheer force. Pandemics and climate changes are powerful. They dwarf us, reminding us of our tiny place within infinite universes.
>
> Those of us drawn to sound know vibration as humbling and infinite, as well. Frequencies have a deep taproot in our collective memories. Perhaps this is why so many of us are attracted to sound. We simply *know* there is a common thread that binds us all together like a social cohesion past-life memory!
>
> In these days, we're facing an uncertain world mired in disruption. *Where do sound and soundworkers fit in this chaotic state?*
>
> We know historically that music has a proven role in stressed cultures. It reminds us of harmony, it brings

us hope, it emotionally binds us together. This is social cohesion. Music provides courage and inspiration. Previously, I simply thought of music performance and audience. Now, I think of nutrient sound as a balm for my neighbors. . . [and I feel] the nutrients of sound are as important to the nervous system as food is for the digestive system.

This is our soundwork now: bringing the nutrients of sound to distressed communities—not for entertainment, but for calming, reassuring.

The challenge is at least twofold:

Finding the therapeutic applications of tone.

Adapting to a much larger context than a handful or hundreds.

Just as our focus turns to the sculpting of therapeutic sound, the concept of audience changes from passive receiver to active participants.

The challenge for the coming age is flexibility and acceptance. These words are easy to read on a page, but not so easy to live. The challenge for the vibroacoustic soundworker is to fully accept that this is a new time, unlike anything most of us have ever experienced.

As a soundworker during times of loss and change, is it enough to bring an hour of relaxation or an individual psychological breakthrough?

As a soundworker, perhaps it is about teaching as many children as possible the art and craft of playing tone, including purposes and results, and then sending these children back to their families as the youngest healers.

Perhaps, it is about the use of the voice in addition to the other instruments of tone and about neighborhood training in these arts, as well?

Be it the Tibetan long horns, the voices of children, or daily sound therapy chill sessions in neighborhood community centers, there is much to consider about the tones of frequencies in days of community need.

Vision

In one man's humble opinion, there is a lot of noise out there. I think it is going to get louder as elections come and go, as food and water become more costly, as individual rights and public health collide, as wars wreak havoc, and more definitively, as our Mother Earth changes in response to our bidding.

And what do we do, as soundworkers who are committed to bringing love and healing, now that the audience has changed in size and need?

How do we use pure tone, raw or refined, to address social cohesion? We must listen deeply to study and create new pathways.

This is a great time. Beyond the doom and the gloom, a new consciousness will emerge. It will cause each of us to go to places vastly different from where we have ever gone before. Challenging times? Yes. But it also feels like an Onwards to the Renaissance time to me.

You in?

We sure are, Joshua, and that is one of the reasons this book was written! Gratitude, dear friends and respected colleagues, for sharing your insights here and additionally, gratitude to the many sound workers known and unknown around the globe. Thank you all for the hope you bring; for your vision, commitment, and service to create a better world through sacred vibration, sound therapy, and music medicine.

CHAPTER 18

TRUTH

The Music Is Love

"Real isn't how you are made," said the Skin Horse. "It's a thing that happens to you. When a child loves you for a long, long time, not just to play with, but REALLY loves you, then you become Real."

"Does it hurt?" asked the Rabbit.

"Sometimes," said the Skin Horse, for he was always truthful. "When you are Real you don't mind being hurt."

"Does it happen all at once, like being wound up," he asked, "or bit by bit?"

"It doesn't happen all at once," said the Skin Horse. "You become. It takes a long time. That's why it doesn't happen often to people who break easily, or have sharp edges, or who have to be carefully kept. Generally, by the time you are Real, most of your hair has been loved off, and your eyes drop out and you get loose in the joints and very shabby. But these things don't matter at all, because once you are Real you can't be ugly, except to people who don't understand."

— MARGERY WILLIAMS BIANCO, *THE VELVETEEN RABBIT*

In mid-December 2021, Dylan delivered to me a strong message. I was in the Crystal Cadence Studio in bliss, surrounded by hundreds of singing bowls, each a unique, beloved friend. I felt calm and at peace. I had meditated, expressed gratitude, set intentions, and cultivated Love. For the alchemy of Love is the vibrational frequency I have learned to choose. At all times and in all situations. Sometimes it may take me some moments or longer when the situation is challenging, but sound has taught me how to tune my heart and find the vibrations of Love. It is how we empower ourselves; how we embrace, forgive, and let go of the past; set healthy boundaries; release triggers; and create our future. It is how we ground in health, well-being, and wholeness.

On this day, Dylan's voice was loud, and his request was undeniable. "Mom, make the oracle deck now! Mom! There is no time to delay!"

This startled me. For some time, I had been considering creating a sound-healing oracle deck to share my experiences of how crystal singing bowls can transform our lives. The Sacred Science of Sound platform was well established, and my schedule was full of projects, teaching, and trainings. I was giving talks and performing sound-healing concerts, leading crystal singing bowl meditations, and providing consultations with clients. The card deck was on my to-do list with no committed time frame.

But, as this big Angel operates, Dylan had a plan that I could not yet understand. Rather quickly, and motivated by his urging, I began to create. On Christmas Eve we recorded the sounds for the 48 cards and on January 9, 2022, which would have been his 26th birthday, the first draft of the oracle card deck was complete. We explored self-publishing and then through grace, an unexpected door opened. The way was revealed, effortlessly and naturally. Hay House invited me to become part of their family of authors, creating the first-of-its-kind interactive Crystal Sound Healing Oracle deck, which was released with great success in May 2023.

Thanks to my big Angel and the support of our audio engineer Alex Pratsyuk, we created a unique product that included high-fidelity audio recordings of the singing bowls accessible though QR codes. The deck is an integrative way to bring sound wisdom to everyone, anywhere in the world, without the necessity of owning a singing bowl. The Los Angeles–based artist Suz Born did the beautiful artwork integrating her original watercolors, golden sacred geometry, and meaningful photographs. The stunning visuals she lovingly created make the tones of the bowls a multisensory experience. As Big Sean wrote, "Jeralyn, this is a game changer." Yes, Sean, it is, activating the transformational power of sound right at your fingertips. That is our vision!

That boy! The miracle of Dylan's communication never ceases to amaze me. On January 9, 2023—Dylan's 27th birthday—I awoke in an unexpected sadness. The tears were flowing silently in a stream I could not stop. I went to workout, and the tears continued to flow. I exercised in silence, and at the end of my workout, my trainer gave me a long hug. There were no words. The tears rolled down my cheeks all the way home, and I could not hold them back. This unexpected sadness softly cleansed and purified me. As I got out of my car, I looked up. The sun was shining brightly, and the blue sky was laced with big, puffy clouds. I took my house key out of my purse and noticed there was a small package waiting at my door. It was addressed to me and sent by the DHL courier company. I had no idea what it was. Then I saw the return address was that of Hay House. It contained two samples of the oracle card decks that are dedicated to my son, Dylan Sage. Right on a heavenly schedule! They had not been expected until March, but of all days to deliver, they had arrived on January 9, as though they were Dylan's confirmation to me that all was well. "Mom, we did it! Happy 27th birthday to me!"

I wept. I was so stunned and in awe that I could not yet touch the decks. I would pass by them during the day, just looking. And the tears would roll again. It took me until the evening to dare to open the first box. There was a pristine energy present beyond this realm. Sacred. I gently shuffled the cards, set an intention, and drew my first card. Once again, I could not believe what I held in my hands. Of all the 48 cards I could have picked, the first one I drew was number 34, Dylan's Card, Saint Germain—Love beyond Time and Space. I could not have made this up and

I noticed the tears had stopped flowing. The heavenly support amazes me. It is an inexplicable sign of Love for which I am immensely grateful.

The energy was magical as we were creating the oracle deck. At one point while I was organizing which bowls to include, I had to choose between two particular bowls: a blue one and a yellow one. While I was recording the bowl sounds with Alex, he asked me to keep the blue one in and remove the yellow one. I did. Many months later, when he held the first sample deck in his hands, I invited him to choose a card. He shuffled, and out of all 48 cards, he chose card number nine—the card that represented the blue bowl! It had a message for him, embedded more than a year earlier, of Wisdom, Regeneration, and Replenishing—exactly what he needed to hear. We were amazed. I find great delight and wonder in all the synchronicities that continue to show up as people around the world use the oracle deck.

Dylan's passing and subsequent guidance has led to another bigger purpose: a book delving deep into crystalline sound and music as medicine. And now you have this precious and powerful book in your hands, containing the interweaving of a life immersed in music, an intensely personal journey, and a practical guide to the crystal singing bowls, one of the most important sound-healing instruments of our time.

The quotation that begins this chapter is one I have loved since childhood. It speaks to the power of authenticity and what truly matters. I understand it now in a deeper way as an adult with so much life experience behind me. An education, a fulfilling career, a family, unexpected loss. Incomprehensible grief. All of it leading to this moment. Learning to breathe, feel, and accept. Letting go of discordance. Choosing harmony. Recognizing the magnitude of being able and willing to love all that is. Recalibrating. Re-Tuning. Learning to trust in the perfection of it all.

Perfection? Yes. Every New Year's Eve for the past few years, Dylan has brought me a strong message.

The first was in 2019. "Mom, it has begun. Fasten your seat belt. Away we go!"

The second was in 2020. "Mom! We're in for a ride, buckle up!— tempo increasing."

The third was in 2021. "Mom! Strap in, we're on a rocket ship!"

The fourth year, New Year's Eve 2022, it completely shifted: In 2023, the year the oracle cards would publish and the book would be finished, with a scheduled release in summer 2024, there was a remarkable change in tone. A softness. A calmness. I gazed into the vastness of the night sky December 31, 2022. I was transported out among the stars into those galaxies that Dylan and I so loved. I heard his voice, tender and reverent: "Mom" he said, "we've arrived. Enjoy the view. Enjoy your life. Mom, we have arrived!"

The energy was simple. Reflective. Every journey has its length, its time needed, its distance, and its characteristics, all of which leave their indelible mark on us. And then we arrive at the destination. It was time to enjoy the view. To be together in this miraculous way.

When I think about the beginning of everything being sound vibration, and I link that thread of creation back to my own childhood, the bigger picture all makes sense: hearing the message at age four that music and singing would be my life's path; having parents who supported me in my visions and trusted me; my faith and belief in God. These three elements were the pillars that shaped my life so that when the unexpected happened, and I had to take the fork in the road, there was a foundation built on sound and love that helped me alchemize the experience. Part of my healing has come from an embrace of words—and their vibrations—that have meaning to me. They guide me like a song or mantra. Words like "mom," "trust," "I love you," "forgiveness," "gratitude," "wholeness," "eternal." Meaningful words that keep resonating and echoing in the silence. When our hearts are shattered, how do we knit ourselves back together? How do we honor and live the uniqueness of our personal alchemy no matter what life gives us? Can we recognize our power as an alchemist to transmute anything? Alchemizing is the process of turning lead into gold. The Japanese art of kintsugi repairs broken pottery with gold, making the refurbished piece even more beautiful than the original. This is a metaphor for our spiritual evolution. When we cannot comprehend our role in the ever-changing symphony of life, it is enough to understand that we are born with our own distinctive frequency, our unique tone, and that nothing can give it to us or take it away from us. It is born in the stillness of Creation, in the anahata, the space of silence in our heart. It is everything we are and everything we will become. All we must do is give it permission to express itself without judgment.

THE EVOLUTION OF CRYSTALLINE SOUND

Today, the crystalline instruments are being featured in music of all genres, from Jhené Aiko's Grammy-nominated album *Chilombo* and her modern mantra, to Ashana's angelically voiced songs and relaxing meditations, to musical mystic India Arie and her "songversations," and to the *Balm* album I played with Jahnavi Harrison. Jahnavi includes the bowls in her live events featuring kirtan and pop-style spiritual ballads. Kevin James and Lee Harris play them in their music, and Devi Brown, a master well-being educator and multidisciplinary healer, integrates them in her meditations and her *Deeply Well* podcast. Musician and producer Maejor featured singing bowls in his song called, "Free, 432 Hz," which you can hear on our app, Source. They are also included in the popular sleep-music series *Sleep Soul* lullabies (I play the bowls on the track "Healing Sounds & Frequencies for Children," *Sleep Soul* Volume 2), and they are included on apps such as Calm, Insight Timer, and the Sacred Science of Sound app, Source. They are integrated into classical music in symphony and opera. Music and the way we listen to it is evolving to meet our need for inner peace and stability amid so many outside influences and distractions. There is fluidity and presence in the soothing vibrations of the crystal instruments. They are a leading player in the world of healing sound.

WHAT IS YOUR VISION FOR YOURSELF?

And now, I invite you to consider the question: what is my vision for myself? How do I embrace my past and create my present and my desired future?

If you are not already doing so, can you now imagine expressing your unique vibrational signature and true purpose, living your life fully present, inspired, abundant, and fulfilled?

My vision of sound medicine is you freely playing your unique instrument, your voice—be it hummed, spoken, toned, chanted, or sung. Perhaps you choose to entrain your voice with the crystalline instruments. There are two-hour loops of pure alchemy sound vibration on our YouTube channel, Crystal Cadence, and I invite you to explore them. Whatever your personal choice of sound or music is, may it help you through any moments of feeling lost or hopeless, helpless, frustrated, or angry, as well as celebrating life and all its blessings. The transformative power of sound is available to all and floats us gently into the realm of Spirit, of the Essential, of the Soul, where we can experience the intangible alchemy of Love.

A NEW PATH WITH MUSIC AS THE MEDICINE

Great music ignites memories that accompany us throughout our life's journey. Among others, Mozart, Beethoven, and Gershwin have been prominent composers in my life. Their brilliant gifts of melody and harmonic structure shaped me. They are musical geniuses who pushed the boundaries of the arts in their lifetimes. It is said that Mozart has sold more albums than Beyoncé and is one of the three most-performed composers of all time. He was a master of incredible melodies who excelled in all musical formats. And he rocked on the keyboard and violin!

Before Beethoven unleashed his genius into his Ninth Symphony, no one had ever written a symphony like that before, and he infused into the extraordinary music the passion of artistry and the anguish and glory of the human heart. It was revolutionary to include a fourth movement with full choir and four soloists. I was privileged not only to sing his Ninth Symphony but also numerous productions of his opera, *Fidelio*. Gershwin's 1924 classic composition "Rhapsody in Blue," with its famous opening clarinet glissando, is perhaps one of the most popular masterpieces of the 20th century. An incredible rendition of it took place during the 1984 Olympics in Los Angeles, when 84 grand pianos converged at the opening ceremonies in an unforgettable array of musical splendor. Mozart, Beethoven, and Gershwin: all three of these brilliant composers changed the field of music in their lifetimes, and all when they were relatively young.

Sometimes I ask myself: if Mozart, Beethoven, and Gershwin were alive today, would they still be stretching boundaries—perhaps prolifically scoring films and creating social-media music clips and video games alike? They were great artists and innovators who advanced humanity through their choice of musical structure, harmonics, melody, and rhythms. Would they still be musical geniuses, creating new genres combining classical music structure with meditation, hip hop or rap? Perhaps Beethoven would have scored his Tenth Symphony for orchestra and crystal bowls, including a rhythm section, a massive choir, soloists who rapped; and composed a vocal section based on a poem by Amanda Gorman, the youngest inaugural poet in American history. Would Mozart have written a series of chamber-orchestra meditations scored for crystal bowls, didgeridoo, and strings? Might he have integrated whale and dolphin sounds mixed with Lemurian Seed Crystal singing bowls in a symphonic work that included gongs? Would he have added tuning forks to a quartet? Perhaps Gershwin and Lin-Manuel Miranda (the composer of the extraordinary, genre-breaking musical *Hamilton*) would have collaborated to create something else that has never been done before: a sound-healing musical created in a "modern mantra," rap-like, jazz-pop style. It's fun to imagine how these musical giants might have integrated sound-healing instruments into their compositions and what they might have created in our time.

CELEBRATING NINE, THE NINTH, AND THE POWER OF GREAT MUSIC

I remember the profound physical sensations of singing nine performances of the Ninth Symphony with Dylan in my womb, and I reflect on the fact that he was born on the ninth of January, two weeks earlier than expected. The number 999 symbolizes completion and destiny. He and I are intrinsically bound together, in life and in death. It was a perfect gesture of the universe for us both to be a part of the genius of Beethoven and his last symphonic masterpiece. That music always calmed my son. He knew it well on a cellular level. He had been vibrating to it in my womb.

"Ode to Joy," taken from Friedrich Schiller's original German poem "An die Freude," is part of the fourth and final movement of the Ninth Symphony. It has been played around the world, and there are pop renditions of the famous melody—including Carrie Underwood's "Joyful, Joyful, We Adore Thee" and the rousing version from the Broadway musical *Sister Act*. In 1985 Beethoven's "Ode to Joy" became the official anthem of the European Union, with its vibrational sentiment of all people living in Peace and Harmony. For many, it is the ultimate symphonic composition, undeniably, one of the greatest pieces of classical music ever composed. I was inspired to use crystal singing bowls to create a version of this melody with musicians Chris Sholar, Geoffro (Geoff Earley), and special guests to coincide with the release of this book. I've translated a portion of Schiller's poem into English, and this loose translation expresses my take on the essence of the last movement of Beethoven's masterpiece.

> *The Creator dwells above, in the celestial realm,*
> *Somewhere beyond the stars,*
> *Where heavenly feathered wings gently flutter,*
> *A place of safety, a haven for all Humankind.*
> *Joy! We are One.*
> *Creating Magic. Living our Wholeness.*
> *Unity through Music*
> *Light. Spark of the Divine.*

I remember Dylan's visitation and the tiny white feather he left intertwined in the threads of my nightgown. *Feathered wings, gently fluttering, somewhere beyond the stars.* He too dwells above the stars now. With the Creator. Among the angels.

What will our futures bring? We cannot know; yet we can decide to seek joy in all things.

My beloved child taught me so much—from the time he was conceived, to his birth, throughout his earthly life, and especially now, in his death. He continues to teach me from his current place as a part of all that is. How my heart still longs to create a future with him, watching as he grows into manhood, walking his life path, whatever that would have been. He reminds me of his destiny: "Mom, this is it, me and you! Bridging the celestial and earthly realms. Not in this human form, Mom, but in Spirit, in a Divine and Eternal partnership." How I love him in whatever form he is. I am reminded of how we used to write each other postcards regularly when I was traveling for singing engagements in Europe. Once, when Dylan was 10 years old, I arrived home from concerts in France, and he was waiting for me at the front door. He was so sweet. He hugged and kissed me and gave me a gift and what I thought then was a very creative and endearing postcard. It was a beautiful photo of the sunlight emerging through the clouds over the ocean with a comment bubble (he loved comics!) that said, "Hi Mom." On the back of the postcard, he wrote these words:

Dear Mom, I love you and I'm so glad to see you again. I've missed you very much. Just remember, always let God's light shine on you and you'll see me!

Much, much, much love from your son Dylan.

Truth

Dylan's comic-strip bubble on the front of the postcard he gave me.

The back of the postcard from my 10-year-old son.

Five years later when we had left Germany and moved home to California, he took a photograph similar to the one on the postcard, printed it, and gifted it to me. Did he sense this place beyond the clouds would soon be his home? Some months after he died, I was going through our boxes of memories, and I found his postcard. I wept when I read it, as I had forgotten what he had written to me. Did he know at age 10 that his life would be short and he would be leaving before me? Was he writing me words of comfort back then to guide me after he departed? We both trusted our yes. I was his mother; he was my son, and he became my greatest teacher.

The Light shines through the clouds, and there is God. (Photo by Dylan)

And he is still teaching me to trust my vulnerability and to speak into the unspeakable as I learn to embody the wisdom of both Life and Death, the Invisible and the Visible. It is our responsibility to rise and embrace the purpose of our lives with a committed *yes*. It is ours to be stable, and immutable in Love. I now know that when we have the courage to lean in to our sorrows, breathe, feel, and accept, we become the vibration of joy. I would not have believed this earlier in my life, but it is true. I see it in my eyes. Sound vibration transformed darkness into an inner radiance. Dylan is always with me, imprinted in my heart. We must trust that everything in our lives is the unfolding of our highest expression of good. So much of what has been revealed through sound frequencies has opened me to understand a deeper life purpose. I am forever grateful.

From my heart to yours, I set an intention for you: wherever you may be in your life, whatever may be your struggles, challenges, delights, and celebrations, may you experience how sacred vibrations, crystalline sound, and healing music can uplift and transform your life.

There, above the stars, beyond the seen, in the place of unity consciousness, pure awareness, frequency and vibration; the invisible, the ineffable, there, where we come from, and there where we will return, lies a connection, a bridge, and that bridge is us, the Light in human form. And the music? Play your human instrument. Fully. The Music is Love.

I wish you Peace.

A FINAL WORD

Regardless of where you are on your journey with the extraordinary vibrations of the crystal singing bowls, I encourage you to take the time to explore and integrate them into your daily life. You can access many sonic crystal experiences with me through the free library of meditations on the Crystal Cadence by Jeralyn Glass YouTube channel, the Source app (where you will also meet some of the sound wisdom contributors to this book), and the first-of-its-kind interactive *Crystal Sound Healing Oracle* card deck. Additional music is also available through crystalcadence.com/new-music and all streaming platforms. I especially invite you to enjoy *Pentatonic Visions: A Journey Through Crystalline Landscapes*, the new album I've created with Chris Sholar to accompany this book, and a rendition of Beethoven's "Ode to Joy" arranged by Geoffro (Geoff Earley) and Chris Sholar with crystal singing bowls and special guests. Coming soon, too, is a new album produced and arranged by Perry LaMarca that revisits great classic melodies in combination with vocals and crystal singing bowls.

And please remember to enjoy the downloadable meditations and short film available by scanning the QR code below or visiting **crystalcadence.com/sacred-vibrations-content**.

There you will find the following downloads:

Meditation #1: The Pentatonic Scale @ 432 Hz

Meditation #2: The Pentatonic Scale @ 528 Hz

Meditation #3: Supergrades Shimmering Beat Frequency

Meditation #4: The Key of C @ 440 Hz

Meditation #5: "Vibrance" from the Album Vibrance @ 440 Hz

Meditation #6: Sacred Vibrations Short Film

Additionally, I encourage you to visit me on my websites **crystalcadence.com, jeralynglass.com, sacredsciencesound.com**, and through social media @crystalcadencela. You may wish to join me in our members' portal, accessible through **https://crystalcadence.com/members/**.

Finally, I invite you to find me on the apps Insight Timer and Source, and on Humanity Streams, Apple Music, Spotify, and all other streaming platforms.

Find out more about my friends and collaborators here:

Healing Sounds: Jonathan and Andi Goldman, healingsounds.com

Sound Made Visible: John Stuart Reid, cymascope.com

Biofield Tuning: Eileen McKusick, biofieldtuning.com

Morter Institute for Bio-energetics: Dr. Sue Morter, drsuemorter.com

Black String Triage Ensemble and Dayvin Hallmon: theblackstringtriageensemble.org

Minerva University: minerva.edu

Joshua Leeds: joshualeeds.com

Dr. Daniel J. Levitin: daniellevitin.com

Victor Wooten: victorwooten.com and vixcamps.com

Jahnavi Harrison: jahnavimusic.com

Anders Holte and Cacina Meadu: anders-holte.com

A Final Word

Suz Born, images for this book and artist for the *Crystal Sound Healing Oracle*: suzborn.com
Kevin James: kevinjamesmusic.com
Jhené Aiko: gotoheal.com and jheneaiko.com
Sean Anderson: uknowbigsean.com
Lauren London: @laurenlondon
Dr. John Beaulieu: biosonics.com

My thanks go to everyone whose crystal alchemy singing bowl sets I have been privileged to curate. It is an honor to put these one-of-a-kind healing instruments together for you.

To all of the students of the Sacred Science of Sound Trainings, and to our Members Portal community, I thank you for allowing me to educate and inspire you with sacred vibrations.

For the committed practitioners—all the graduates of the Sacred Science of Sound Trainings and especially to those who shared their experiences in this book: I am filled with gratitude for all you are and all you do! Your gifts of Sacred Vibrations are changing the world.

Wendy Leppard: The Space in Between, thespaceinbetween.co.za, South Africa
Amy Bacheller: Scent from Heaven, California, scentfromheaven-sb.com
Dr. Maja Jurisic, MD: Retune, Refresh, Restore, resoundinglife.net, Wisconsin
Sara Bayles: Soul Wave Wellness, California, soulwavewellness.com
Dr. Sole Carbone: inner-tuning.com Nova Scotia
Melissa Paddison: aligntorise.org, @aligntorise, Australia
Laura Solveig Israel: lauraisrael.de, Germany
Susan Eva: susaneva.com, Canada
Brenda Cochran: iaminlight.com, Oregon
Nada Hogan: nadahogan.com, Minnesota
Dr. Greg Eckel: energy4lifecenters.com, Utah
Angela Weisman: soundtruthllc.com, Ohio

Jessica Niedeffer: adaracollective.org and agadaenergyhealing.com, Texas

Anne Johnson: alexandertechphiladelphia.com, Pennsylvania

In Love, Light, and Healing Sound,
Jeralyn Glass

ENDNOTES

CHAPTER 1

1. Gerald Sinclair, "Scientists Prove That Everything Is Energy—Is Reality Real?," Awareness Act, April 21, 2019, accessed September 11, 2023, https://awarenessact.com/scientists-prove-that-everything-is-energy-is-reality-real/.

2. Lev S. Tsimring, "Noise in Biology," *Reports on Progress in Physics* 77, no. 2 (February 2014): 026601, https://doi.org/10.1088/0034-4885/77/2/026601.

3. Sam Wong, "Your Bones Contain Crystals Shaped Like Fingers and Hands," *New Scientist*, May 3, 2018, accessed September 11, 2023, https://www.newscientist.com/article/2168058-your-bones-contain-crystals-shaped-like-fingers-and-hands/; Richard Templer and John Seddon, "The World of Liquid Crystals," *New Scientist*, May 18, 1991, accessed September 11, 2023, https://www.newscientist.com/article/mg13017695-400-the-world-of-liquid-crystals/.

4. Mona Lisa Chanda and Daniel J. Levitin, "The Neurochemistry of Music," *Trends in Cognitive Sciences* 17, no. 4 (April 2013): 179–193, https://doi.org/10.1016/j.tics.2013.02.007.

5. Jane M. Simoni, Ph.D., "Sound Science," National Institutes of Health, January 23, 2024, accessed March 26, 2024, https://obssr.od.nih.gov/news-and-events/news/director-voice/sound-science-conversation-drs-simoni-and-langevin.

6. Saifman, J., Colverson, A., Prem, A., Chomiak, J., & Doré, S. (2023). Therapeutic Potential of Music-Based Interventions on the Stress Response and Neuroinflammatory Biomarkers in COVID-19: A Review. *Music & Science, 6*, 20592043221135808.

7. Bissonnette, J., Dumont, E., Pinard, A. M., Landry, M., Rainville, P., & Ogez, D. (2023). Hypnosis and music interventions for anxiety, pain, sleep and well-being in palliative care: systematic review and meta-analysis. *BMJ supportive & palliative care, 13*(e3), e503-e514.

8. Jonathan Goldman and Andi Goldman, *The Humming Effect: Sound Healing for Health and Happiness* (Rochester, VT: Healing Arts Press, 2017).

CHAPTER 4

1. "Cymatics Research—The Physics of Sound," Cymascope, accessed September 11, 2023, https://cymascope.com/the-physics-of-sound/; "Sound Therapy—201: Biological Mechanisms," Cymascope, accessed September 11, 2023, https://cymascope.com/sound-therapy-201/.

CHAPTER 6

1. Reznikov et al., "Fractal-like Hierarchical Organization of Bone Begins at the Nanoscale," *Science* 36, no. 6366 (May 4, 2018): eaao2189, https://doi.org/10.1126/science.aao2189.
2. S. Trokel, "The Physical Basis for Transparency of the Crystalline Lens," *Investigative Ophthalmology* 1 (August 1962): 493–501, PMID: 13922578.
3. Pelling et al., "Local Nanomechanical Motion of the Cell Wall of *Saccharomyces cerevisiae*," *Science* 305, no. 5687 (August 20, 2004): 1147–1150, https://doi.org/10.1126/science.1097640.

CHAPTER 9

1. Sam Roberts, "Mitchell L. Gaynor, 59, Dies; Oncologist and Author on Alternative Treatments," *New York Times*, September 18, 2015.
2. Mitchell L. Gaynor, *The Healing Power of Sound: Recovery from Life-Threatening Illness Using Sound, Voice, and Music* (Boulder, CO: Shambhala Publications, 1999), 191.
3. Hanae Armitage, "Sound Research," *Stanford Medicine Magazine*, May 21, 2018, https://stanmed.stanford.edu/innovations-helping-harness-sound-acoustics-healing/; "Biomagnetic Academy of Spiritual Science (BASS)—Tuning In pt. 2.—Energy Sound Cymatics to Heal," BiomagHealer, January 12, 2022, https://www.youtube.com/watch?v=yCxjl2or1H4.

CHAPTER 12

1. James H. Austin, *Zen and the Brain: Toward an Understanding of Meditation and Consciousness* (Cambridge, MA: MIT Press, 1999); Richard J. Davidson and Antoine Lutz, "Buddha's Brain: Neuroplasticity and Meditation," IEEE Signal Processing Magazine 25, no. 1 (January 1, 2008): 176–174; https://doi.org/10.1109/MSP.2008.4431873.

CHAPTER 14

1. Swami Satyananda Saraswati, *Asana pranayama mudra bandha* (Bihar, India: Bihar School of Yoga, 2008), 531.

CHAPTER 17

1. The Black String Triage Ensemble, https://www.theblackstringtriageensemble.org/mission.
2. Pelling et al., "Local Nanomechanical Motion of the Cell Wall of *Saccharomyces cerevisiae*," *Science* 305, no. 5687 (August 20, 2004): 1147–1150, https://doi.org/10.1126/science.1097640.
3. Reid et al., "Imaging Cancer and Healthy Cell Sounds in Water by Cymascope, Followed by Quantitative Analysis by Planck-Shannon Classifier," *Water* 11 (2020), https://doi.org/10.14294/WATER.2019.6.

GLOSSARY

There are some important concepts to explore as our understanding of sound healing with the crystal singing bowls deepens.

B

A **beat frequency** is the difference in frequency of two sound waves; i.e., when you play two slightly different tunings of bowls with the same note, or you play a half step in music. This creates a beat frequency. I like to call these "shimmering sounds."

A **binaural beat** is the third tone created by the brain when we listen to two distinct tones through headphones at different frequencies in different ears. When you listen to two slightly different frequencies, you will hear the difference in frequency between the left and right ear. For example, if the left ear registers a tone at 220 Hz and the right at 215 Hz, the binaural beat heard is the difference between the two frequencies—5 Hz—which will entrain the brain into the theta brain-wave state.

Brain-wave states are specific frequencies produced in the brain. The most common ones are gamma, delta, theta, alpha, and beta. Each has its own characteristics that correspond with different physiological, mental, and emotional processes.

E

Entrainment is when two objects come in contact with each other and, amazingly, the vibrational frequencies eventually fall into sync. This phenomenon exists for the purpose of conserving energy. Rather than fighting each other's unique resonance, the stronger frequency becomes dominant, and the two objects synchronize. An example of this is when we place a row of metronomes together and start them at different times. Within a very short time, all the metronomes will be swinging together, synchronized in rhythm and in harmony. Entrainment is a concept used often in sound therapy.

F

Frequency is the number of sound waves that go by us or into us per second. It is measured in hertz (Hz), with one hertz equal to one wave (or cycle) per second.

H

Harmony is when two or more frequencies of sound (notes) come together with a pleasing, coherent, harmonic effect.

Hertz (Hz) is a measurement of frequency; i.e., how many sound waves per second. 1 Hz = 1 wave or cycle per second.

I

An **interval** is the distance between two notes, or the number of notes on the scale between two specific notes. (The keyboard graphic on page 166 shows the location of the notes on the scale.) There are certain intervals that make up the structure of an effective sound bath, and they are crucial to setting up parameters of safety. These create a musical structure that allows easy and graceful entry into the subconscious mind, where patterns can reveal and be healed, felt, and transformed. The following are examples of healing intervals:

- A **half step** is the closest note to the neighboring note on a piano keyboard—for example, a white note to the black note directly next to it, or a white note next to the neighboring white note (only where there is no black note in between): i.e., C to C♯ (white to a black or sharp note) or E to F (two white notes).

- A **perfect fourth** is an interval made up of five half steps (as in the song "Here Comes the Bride"). This interval creates stability yet lightness, openness, acceptance, and inner peace. It has a 4/3 frequency ratio, meaning the upper note makes four vibrations or waves in the same amount of time that the lower note makes three.

- A **perfect fifth** is an interval made up of seven half steps (as in the song "Twinkle, Twinkle Little Star"). This is the interval that is used when tuning a piano. It creates a relaxation response and a gateway to higher consciousness, and it is at the core of Pythagorean wisdom in the structure of sound. It has a 3/2 frequency ratio, meaning the upper note makes three vibrations or waves in the same amount of time that the lower note makes two.

- An **octave** is an interval made up of twelve half steps (as in the song "Over the Rainbow"). This interval creates a sound of unity, of being on the same "wavelength." It has a 2/1 frequency ratio, meaning the A note to which orchestras tune today as 440 Hz makes one vibration or wave in the same amount of time that the A note above it (which has 880 Hz) makes two.

P

Pitch is frequency expressed by a letter (named as a musical note such as C, D, E, or F).

R

Resonance is the basic principle of sound healing and relates to the concept that every object has a vibratory frequency, including the entire universe. Everything is in a state of vibration, including human beings. Every organ, cell, bone, tissue, and liquid of the body, and the electromagnetic fields that surround the body, has a healthy vibratory frequency. If we are not resonating with some part of ourselves or our surroundings, we become dissonant and therefore unhealthy; our naturally healthy frequency becomes a frequency that vibrates without harmony, creating illness and dis-ease.

T

Tuning is the vibrational frequency of an instrument. Most instruments can be tuned to a variety of vibrational frequencies, but a crystal alchemy singing bowl is manufactured to have one specific tuning. Some musicians and theorists believe that 432 Hz tuning and 528 Hz tuning have strong positive effects on the human body. In my experience all tunings are positive and an individual preference—there are no scientific studies that confirm that any one tuning has more benefits than another. The three most common tunings are:

– 432 Hz: a frequency that some people believe is connected to nature. A popular tuning in Baroque music and in sound healing.

– 440 Hz: "concert pitch." This is the standard tuning for modern-day music based on the A note.

– 528 Hz: technically the A note at 444 Hz. *528 Hz* refers to the corresponding C note in octave 5.

ACKNOWLEDGMENTS

My deepest gratitude to:

My beloved son, Dylan Sage. I would not be who I have become without you. Your courage, devotion, and constant presence sustain me.

My mom and dad. Thank you for all you have given me! What a wonderful life filled with adventure, learning, and Love we have shared!

To my Grandma Ree Dear and Grandpa Jack: thank you for all the picnics, for your seahorse wisdom, and for your love.

To my great-grandma, affectionately known as "Nana," your love mattered!

My family: my sisters and brother, my nephews, and their wives and children. I love you to the moon and back.

Annette Warren Smith, the extraordinary voice teacher I began training with when I was 11 years old. You planted seeds in me that have grown and served me my entire life.

Dr. Sue Morter. There is Love bigger than I ever imagined. Your wisdom and guidance sparked a whole new world of possibilities.

Peter Pearce, I love you dearly. You have been at my side, holding my hand and having my back.

My dear friends and beloved colleagues, thank you for your collaborations and contributions as special guests in this book. I am incredibly grateful for your expertise and insights. You are extraordinary and I honor, respect, and love you: Jhené Aiko, Big Sean, Suz Born, Lalah Delia, Jonathan and Andi Goldman, Jahnavi Harrison, Anders Holte and Cacina Meadu, Kevin James, Joshua Leeds, Dr. Daniel J. Levitin, Lauren London, Eileen McKusick, John Stuart Reid, and Victor Wooten.

To my treasured friends: I give gratitude for your support, encouragement, and inspiration. You light up my world! Thanks to: India Arie, Lois Blumenthal, Devi Brown, Han Feng, Renée Fleming, Kyle Gray, Lee Harris, Janet Leahy, Anita Moorjani, Gwyneth Paltrow, Solana Rowe, Jada Pinkett Smith, Marci Shimoff, John and Karina Stewart, and Tricia Williams.

My incredible team: You are so loved and appreciated! Jennifer Beckett, Ananda Prohs, Kaylie Clark, Luis Revilla, Alex Pratsyuk, Dana Austin, and Hilton Goring.

To dear Susan Crossman, without your wisdom, expertise, loving support of my vision, and commitment to excellence, this book would not have been.

A special thank you to a brilliant Light: beloved friend Dr. Sole Carbone; you understand so deeply and share your wisdom so freely.

Thank you, Dr. Douglas Burnett Smith. You know how to dot your i's and cross your t's!

Young Noah. You bring hope. I am blessed to be your Godmama.

I am grateful to these amazing people, dear friends, guides, and healers who hold me, support me, and inspire me. You make this world a better place by your presence: Janet Glazier, Annette Tixier, Tanya Malott, Barry Goldstein, Deva Premal and Miten, Holiday Higa, Brenda Rew, Bhavna Bhatia, Dr. Kimberly Hoffman, Mary Horst, Dr. Elisa Zinberg, Jadie Brynestadt, DeAnn Zelenkov, Lisa Strogal, Joyce Sharman, Noelle Nese Mercer, Nina Collier, Dr. Christine Rathenow, Helena and Karl-Heinz Urlaub, Mila J, Miyoku Chilombo, Katrina Askew, Lisa Teran, Dr. Bernie Schulte, Sujay Krishna Seshadri, Dr. Eva Maria Schneider, Dr. Allan and Jeanne Peters, Dr. Yusuke Mori, Dr. Kerry Haydel, Dr. Allan E. Sosin, Dr. Karl Maret, Brandi Marshall, AlexSandra Leslie, Suzanne Lawlor, Elain Young, Susie Nelson-Smith, Tryshe Dhevney, Edith Echeverria, Ashana, Pamela Butters, Lee Carroll (Kryon), Monika Muranyi, Gregg Braden, Dr. Bruce Lipton, Dr. Todd Ovokaitys, Dr. Kavita Chinnaiyan, Dr. Kulreet Chaudhary, Dr. Charles Limb, Russ Tarleton, Maria and Isidro Ramos, Irene Ingalls, James van Praagh, Suzanne Giesemann, Stephen Dinan and Josh Wise and The Shift, Dr. Greg and Kitty Eckel-Stoneburner and Energy4Life Center,

Acknowledgments

Kimba Arem, Marianne Williamson, Londrelle, Maejor, the Brothers Koren, Dr. Ibrahim Karim and Doreya Karim, Mike Magee, Vern Falby, Elaine Welteroth, Jonathan Singletary, and Jay Shetty and Radhi Devlukia.

To the Cancer Support Community South Bay, especially Nancy Lomibao: Thank you for giving me a safe place to lead my first sound baths.

To Trinity Care Hospice for holding Sacred Space as you do.

To Paul, Christine, and St. Peter's by the Sea: I am grateful for this place of love and shelter that has been my home since childhood.

Heartfelt gratitude to Susanne Klatten for her generous support of my vision for Kids4Kids and thank you to Mary Hammond, our artistic advisor, Christoph Weinhart, musical director, and Dr. Monika Nöcker-Ribaupierre, head of our music therapy program. Thanks also go to Jeffrey Trinklein, Martin Schmidt, and Gibson Dunn & Crutcher, LLP, Munich. To all the performers who participated in Kids4Kids World Foundation productions in Munich, Germany, from 2006 to 2013.

Heartfelt thanks to Anna Cooperberg, Patty Gift, Julia Pastore, Reid Tracy, and the Hay House team for your wonderful support in the creation of my first book!

With deep reverence for every moment of this human life. How extraordinary it has been, and how blessed and honored I am to share *Sacred Vibrations* with you. Thank you, dear reader, for being a part of this journey.

ABOUT THE AUTHOR

JERALYN GLASS helps people facilitate healing and transformation with the power of music and sacred frequencies that nourish our bodies, connect us with our hearts, and awaken our consciousness. She is the author of *Crystal Sound Healing Oracle*. She is an internationally known, multidisciplinary musician whose career began on Broadway and took her to the opera and concert stages of the world. She founded Crystal Cadence Sound Healing Studio LLC and the Sacred Science of Sound with its app, Source, as educational platforms where science, spirituality, energy medicine, sound vibration, and the healing power of music intersect. She has performed her high-vibrational music alongside some of the most respected New Thought leaders and best-selling authors, and has collaborated with popular musicians. Her music is available at **crystalcadence.com**, the YouTube channel Crystal Cadence by Jeralyn Glass, and on all streaming platforms. Learn more at **jeralynglass.com**.

HAY HOUSE TITLES OF RELATED INTEREST

YOU CAN HEAL YOUR LIFE, the movie,
starring Louise Hay & Friends
(available as an online streaming video)
www.hayhouse.com/louise-movie

THE SHIFT, the movie,
starring Dr. Wayne W. Dyer
(available as an online streaming video)
www.hayhouse.com/the-shift-movie

***

THE 7 SECRETS OF SOUND HEALING by Jonathan Goldman

THETAHEALING™: Introducing an Extraordinary Energy-Healing Modality by Vianna Stibal

THE SCIENCE BEHIND TAPPING: A Proven Stress Management Technique for the Mind and Body by Peta Stapleton, MD

UNBLOCKED: A Revolutionary Approach to Tapping into Your Chakra Empowerment Energy to Reclaim Your Passion, Joy, and Confidence by Margaret Lynch Raniere and David Raniere, PhD

All of the above are available at your local bookstore,
or may be ordered by contacting Hay House.

***

MEDITATE.
VISUALIZE.
LEARN.

Get the **Empower You**
Unlimited Audio *Mobile App*

Get unlimited access to the entire Hay House audio library!

You'll get:

- 500+ inspiring and life-changing **audiobooks**
- 700+ ad-free **guided meditations** for sleep, healing, relaxation, spiritual connection, and more
- Hundreds of audios **under 20 minutes** to easily fit into your day
- **Exclusive content** *only* for subscribers
- No credits, **no limits**

New audios added every week!

 ★★★★★ **I ADORE this app.**
I use it almost every day. Such a blessing. – Aya Lucy Rose

Scan me with your phone camera!

TRY FOR FREE!
Go to: hayhouse.co.uk/listen-free

HAY HOUSE
Online Video Courses

Your journey to a better life starts with figuring out which path is best for you. Hay House Online Courses provide guidance in mental and physical health, personal finance, telling your unique story, and so much more!

LEARN HOW TO:

- choose your words and actions wisely so you can tap into life's magic
- clear the energy in yourself and your environments for improved clarity, peace, and joy
- forgive, visualize, and trust in order to create a life of authenticity and abundance
- manifest lifelong health by improving nutrition, reducing stress, improving sleep, and more
- create your own unique angelic communication toolkit to help you to receive clear messages for yourself and others
- use the creative power of the quantum realm to create health and well-being

To find the guide for your journey, visit www.HayHouseU.com.

HAY HOUSE
online learning

CONNECT WITH
HAY HOUSE
ONLINE

🌐 hayhouse.co.uk   f @hayhouse

◉ @hayhouseuk   𝕏 @hayhouseuk

▶ @hayhouseuk   ♪ @hayhouseuk

Find out all about our latest books & card decks • Be the first to know about exclusive discounts • Interact with our authors in live broadcasts • Celebrate the cycle of the seasons with us • Watch free videos from your favourite authors • Connect with like-minded souls

---

'The gateways to wisdom and knowledge are always open.'

Louise Hay